Climate Change and the Energy Transition

The climate crisis is on the brink of its threshold of no return, and there is great urgency for finding a viable solution to this existential threat. But many have still not woken up to the seriousness of the situation. This book attempts to remove some of the key knowledge barriers and proposes a radically new strategy to treat the crisis.

The book takes the reader on a learning journey through the cause of global warming, the impacts of climate change, and the science behind changes in the Earth's climate system. A detailed introduction to how the Earth's climate system is monitored by NASA, the European Space Agency (ESA) and the US National Oceanic and Atmospheric Administration (NOAA) using projects such as the Clouds and the Earth's Radiant Energy System (CERES) project follows, and how the Intergovernmental Panel on Climate Change (IPCC) teams together with climate modelling centres to predict future climate change using climate models is explained.

The reader is treated with a history lesson on the UN Framework Convention on Climate Change (UNFCCC) and its Conference of the Parties (COPs) as well as the Paris Agreement, the latter of which signals the beginning of an organised effort to tackle the climate issue. The role of renewable energy in the energy transition is elaborated. Attention next dwells on the state of the global climate near 1.5°C. The final chapter proposes a new climate action framework and action strategy, dubbed the *rationalised net-zero strategy*, that employs an innovative process to include external influences on climate action into a new unified strategy.

The book is primarily intended for climate policymakers, NGOs, climate action practitioners and advocates, as well as academics, government and private sector decision-makers and the climate-conscious public at large.

Key Features:

- Written in plain English.
- A comprehensive, coherent and timely treatment of the climate crisis, from the science background to the policies, and the required energy transition and strategies needed to implement them.
- Provides a contiguous historical record of the progress of climate action from the Paris Agreement to the present.

From Atoms to Atmospheres: The Physics of Climate

Other recent books in the series:

Climate Change and the Energy Transition: The Science, Politics and Solutions for a Sustainable Future
Anirudh Singh

Climate Change and the Energy Transition

The Science, Politics, and Solutions
for a Sustainable Future

Anirudh Singh

CRC Press
Taylor & Francis Group
Boca Raton London New York

CRC Press is an imprint of the
Taylor & Francis Group, an **informa** business

Designed cover image: Shutterstock

First edition published 2026
by CRC Press
2385 NW Executive Center Drive, Suite 320, Boca Raton FL 33431

and by CRC Press
4 Park Square, Milton Park, Abingdon, Oxon, OX14 4RN

CRC Press is an imprint of Taylor & Francis Group, LLC

ISBN: 978-1-032-87157-8 (hbk)
ISBN: 978-1-032-86794-6 (pbk)
ISBN: 978-1-003-53118-0 (ebk)

DOI: 10.1201/9781003531180

Typeset in Minion
by KnowledgeWorks Global Ltd.

*To the climate-conscious
public at large –*

---

*We may miss the boat with 1.5°C,
but the socio-economic chaos that looms
beyond may still be averted. Our final
hope lies in human goodwill.*

Contents

Preface

The climate crisis is an existential threat. It is also amongst one of the most intractable problems of the world. The main aim of this book is to examine the action that is currently being taken to mitigate this threat and to investigate whether a new solution to this problem can be found.

In addition to this primary aim, it must also be noted with some concern that while there is widespread interest in the subject, the gravity of the climate crisis is not well appreciated by many. A major contributor to this worrying deficiency is a lack of information flow, both between the extended climate action team dedicated to eliminating the threat and the global community as a whole, and internally within the climate action community itself.

Climate action depends in an integral way on the efforts of climate scientists and agencies who gather and disseminate data on the Earth's climate system using space, atmospheric, land and ocean-based observatories, climate modellers who use climate models to predict future climate, the IPCC which produces regular assessment reports on climate change, climate policymakers who implement the Paris Agreement, as well as climate NGOs, activists and the general public who advocate for climate action. Another important goal of this book therefore is to effectively disseminate knowledge of the work done by the climate scientists, agencies and modellers amongst policymakers, NGOs and climate action advocates generally.

This book aims to raise the base level of understanding of the climate crisis and the climate action being undertaken amongst all who are genuinely concerned about its dire outcomes, with the ultimate goal of identifying a new framework of action to bring the run-away situation under control.

Towards these ends, the book begins by noting the current status of the climate crisis. It then examines the cause of global warming and the science behind its impacts on the Earth's climate system.

A detailed introduction is provided of how the Earth's climate system is monitored using various observatories by agencies such as NASA, ESA, NOAA and projects such as CERES and CCI. This is followed by an equally detailed introduction to climate models and the numerous climate modelling centres around the world. Together with the IPCC, these centres use the data made available by the agencies to predict future emissions and climate impact scenarios within the grand collaboration known as the *World Climate Research Program (WCRP)*.

The history of climate action is traced from the establishment of the UNFCCC to its various *Conferences of the Parties (COPs)* that culminated in the Paris Agreement of 2015, and key articles of this Agreement are discussed. An overview of the emissions reduction commitments, including the *nationally determined contributions (NDCs)*, the *long-term low emission development strategies (LT-LEDS)*, as well as the *net-zero strategies*, adopted to implement the Paris Agreement then follows.

Renewable energy is introduced as a clean energy choice for the energy transition, and a critique carried out of nuclear energy as an alternative clean energy source to renewables. An overview is provided of how renewable energy has been employed in the energy transition pathways of two selected countries.

The future state of the global climate after a 1.5°C warming of the Earth has been crossed is next elaborated/speculated upon. The book ends by proposing a new climate action framework and strategy that employs an innovative method of incorporating external influences on climate action to produce an integrated climate action strategy dubbed the *rationalised net-zero strategy (RNZS)*.

Acknowledgements

Choosing to write on the theme of climate change and the energy transition is always prone to risks, due to its constantly evolving nature. Thus, it was with some trepidation that the theme was suggested to Commissioning Editor Rebecca Hodges-Davies. I was pleasantly surprised to discover her willingness to accept the choice and forward it to the review process.

The book took about a year to write and has remained very relevant to the current status of the subject. My grateful thanks to Rebecca for accepting the suggested theme and indeed suggesting the title as well!

I acknowledge and thank Asha Sinha for her grateful assistance with Figures 1.2, 1.3, 4.1, 4.2, 6.1, 8.1, 9.1 and 9.2. It would have been very difficult without this help.

Finally, as always, I thank my family and friends for their patience and understanding and their sacrifice of quality social time during the progress of this book project.

About the Author

Anirudh Singh, Ph.D., is a researcher/consultant/writer in Renewable Energy and Climate Change Mitigation and a former Professor in Renewable Energy. He is currently an Honorary Professor at the University of Southern Queensland and Adjunct Professor at the University of the South Pacific.

Singh earned his Ph.D. in Condensed Matter Physics from Leicester University, UK, and an M.Sc. in Theoretical Nuclear Physics from Auckland University, New Zealand. He has been working in the area of renewable energy/climate change mitigation for the last 18 years.

During his time at the University of the South Pacific (2006–2016), Anirudh taught in the areas of physics and renewable energy and was project leader for two EU-funded energy capacity-building projects involving a global consortium of universities, Section Editor for the *Handbook of Climate Change Adaptation* (Springer) and Expert Reviewer of the Second and First Order Drafts of the Working Group II contribution to the *IPCC Fifth Assessment Report* (AR5). After joining the University of Fiji in 2017, Professor Singh went on to develop a *Masters in Renewable Energy Management* program for postgraduates.

During his academic career, Anirudh produced 14 book chapters, in excess of 60 scientific papers, and presented research seminars and presentations in 11 countries.

Singh's writing career in the renewable energy/climate change field began with the first edition of the book, *Talking Renewables: A Renewable Energy Primer for Everyone*, with the IOP in 2018. This was followed by the publication of his edited book, *Translating the Paris Agreement into Action in the Pacific* in 2020, and three years later, of *Bioenergy for Power Generation, Transportation and Climate Change Mitigation*. His last book was the second edition of *Talking Renewables* that came out in March

2025. He has also published an expository book in physics titled *Concepts and the Foundations of Physics* in 2021.

Before his career in Renewable Energy, Dr. Singh carried out research in muon implantation studies and condensed matter physics at the Rutherford Appleton Laboratory and Leicester University in the UK.

CHAPTER 1

A World in Crisis

1.1 INTRODUCTION

Over the last several years, the world has been in the grips of a multitude of global crises. These crises have had significant socio-economic impacts, both at the national and international scales.

Global crises are always human issues, and as such, their analysis requires an inter-disciplinary approach. This commonality also allows lessons gained from one crisis to provide insights into the solutions of the others. In the most obvious case, this amounts to identifying features that are shared by all. This can be of great value, especially in the quest for a solution to the climate issue.

The climate crisis is a truly existential threat, not only to humans but also to all living organisms. The current methods for its solution are not working, and a new approach is needed. We need to find a new solution to this seemingly intractable issue, and the other crises provide us with a unique opportunity to learn how to achieve this. It is instructive, therefore, to start by examining all these crises as a whole before proceeding to consider the specific case of the climate crisis.

1.2 THE GLOBAL STATUS

The three crises of most relevance are COVID-19, the geopolitical instability and the climate crisis. Let us consider them one by one.

1.2.1 The COVID-19 Pandemic

COVID-19 (COVID) originated in China in November 2019 and was officially declared a pandemic by the World Health Organization (WHO) on 30 January 2020 [1]. By the time of the official end of COVID as a global

DOI: 10.1201/9781003531180-1

health emergency, the WHO had recorded a cumulative 765 million cases and nearly 7 million deaths due to the virus worldwide.

While COVID-19 was nowhere as severe as the Spanish flu of 1918–1919 (which infected 500 million people and killed 50 million), its socio-economic impact was the worst known of any global pandemic. The lockdowns and social isolation mandates that were imposed to prevent the spread of the virus introduced severe restrictions on the freedom of public movement and access to work, education and social life. They also had highly disruptive impacts on national and global economies. National lockdowns to prevent the entry of the virus into countries severely restricted economic supply chains on which globalised economies depend for their products and services. Local restrictions on public movements and gatherings had devastating impacts on local commerce, trade and industry.

The official end of the COVID emergency was declared by the WHO when, on 5 May 2023, it announced that it was no longer an issue of global concern [2].

There seems to have been no official reason stated explicitly for the declaration. But in all probability, it was motivated by the global economic disaster that mandatory national actions due to the emergency declaration were causing. The declaration resulted in countries effectively dismantling the elaborate system of COVID legislations and mandates they had enacted in order to deal with the issue. Life essentially returned to the normalcy of the pre-COVID era, and economies improved rapidly.

It is important to note that neither national governments nor the international authority (WHO) made any attempt to deny that the virus was still present and actively afflicting the global population after the end of the emergency was announced. Indeed, according to the WHO declaration that ended the emergency,

> "the virus .... is here to stay. It is still killing and still changing. The risk remains of new variants emerging that cause new surges in cases and deaths" [2].

Evidently, the hope was that the COVID-19 pandemic would be "normalised" in manner similar to

---

Lessons we can learn from COVID-19:

- Modern economies are critically dependent on global supply chains.
- A global crisis can be "ended" simply through a decision made by the relevant authority (the WHO in the case of COVID).

---

the perennial flu, and a "new normal" would be established where people accepted this new virus as part of the normal cycle of viral infections.

1.2.2 The Geopolitical Scene

Two examples of the political events that are of relevance are the Russian invasion of Ukraine and the aftermath of the election of Donald Trump as president of the US.

1.2.2.1 *The Invasion on Ukraine*

The Russian invasion of Ukraine on 24 February 2022 came as a complete surprise to all. Most initially mistook it for the border exercise it began as but were horrified when it turned into a full-fledged invasion of Ukraine.

The United Nations General Assembly took urgent action and adopted resolution GA/12407 on 2 March 2022, demanding that Russia immediately end the illegal use of force in Ukraine and withdraw all troops [3]. Sanctions were quickly imposed against Russia [4].

The international response was, however, not quite what had been anticipated. The support for the UN condemnation of the Russian invasion was overwhelming, with 141 voting in favour, 5 (Belarus, North Korea, Eritrea, Russian Federation and Syria) against and 35 abstentions. But soon after the imposition of the sanctions, several countries realised that Russia was an important supplier of their oil and gas needs [5]. These countries consequently chose to compromise their stance in favour of their economic interests, with the result that the efficacy of the sanctions was significantly weakened.

> Lesson: The economic interests of countries can be more important than considerations of the sovereign rights of nations such as Ukraine.

1.2.2.2 *US Politics – Trump and Climate Action*

After his election as President of the US, Donald Trump withdrew the membership of the US from the Paris Agreement [6, 7]. The nation was only re-admitted to the climate treaty after Joe Biden's election as President in the following term.

> Lesson: The national politics of a country can have a determining influence on global climate change policies. As a result, the course of global climate action can be dramatically altered by the election of a national leader with a divergent belief system about climate change.

What are the lessons learnt from these geopolitical crises?

1.2.3 The Climate Crisis

The impacts of global warming have worsened dramatically since the start of this decade. The strength and frequency of severe weather events such as wildfires, heat waves and cold spells, storms and floods are rising exponentially.

The World Meteorological Organization (WMO) has warned that 1.5°C will be breached at least temporarily within the next five years [8]. However, evidence of climate tipping point phenomena, such as polar and glacial ice melts, ocean warming and coral bleaching and other examples of ecosystem collapse, are already being felt worldwide. Meanwhile, annual global emissions are still rising, despite the concerted efforts by the UN Framework Convention on Climate Change (UNFCCC) to contain them through the mechanism of the Paris Agreement.

Taken together, the above geopolitical events show how:

- seemingly disparate issues such as US politics and the climate crisis can be related,
- economic concerns of countries have priority over principles of sovereignty and democratic rights, and
- the solution pathway for the global climate issue can be severely constrained by the political ideologies of leaders.

It is evident that the current methods of addressing climate change are ineffectual, and newer methods must be resorted to. Part of the issue with the Paris Agreement is the limited scope of strategies the Agreement allows in its action plan. In particular, it is totally dependent on the single strategy of reducing net emissions. There is a need for a holistic review of the methodology adopted to mitigate global warming.

How can we understand the climate crisis better by learning from the other crises? We can do this by first noting that the current (if not all) global crises are part of intricate networks of relationships where the global economy invariably serves as a common denominator. The climate crisis is no exception. Secondly, national and international priorities have the ultimate deciding role when a country is confronted with a global crisis. The following two sections elaborate on these observations further and demonstrate how examining the common properties of global crises can provide indispensable insights into finding solutions to the climate crisis.

1.3 GLOBAL CRISES ARE RELATED

Global crises are frequently related to each other, and all are related to the economy. The strong correlation between COVID-19 and climate change was demonstrated graphically during the active duration of the COVID period through a sharp drop in global greenhouse gas (GHG) emissions. This is revealed in Figure 1.1, which shows how the annual global emissions responded to the global lockdowns during the 2019–2023 period.

The pronounced dip in emissions in 2020 is strongly correlated with the sharp decrease in the use of (CO_2 emitting) aviation transportation during that year.

Thus, while the COVID lockdowns adversely affected the global economy, they actually benefitted the cause of climate change mitigation by inadvertently reducing global GHG emissions. This can be expressed by saying that COVID-19 and the economy are **anti-coupled** to each other, while COVID-19 and climate change are **directly coupled.** Such relationships between global crises are expressed pictorially in Figure 1.2.

The above analyses show that the economy is a common theme to all crises. They also reveal that economic necessities over-ride all other issues. This was clearly demonstrated in the decisions made at the start of the Russian invasion of Ukraine by countries importing energy from Russia.

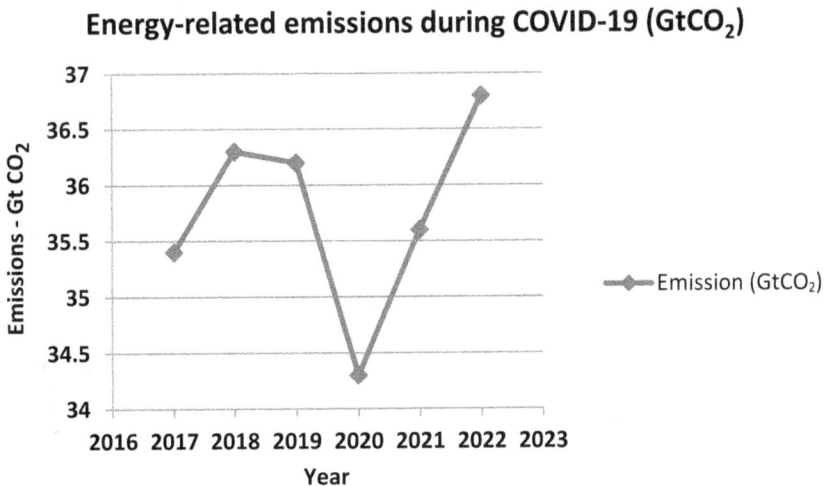

FIGURE 1.1   Graph showing the energy-related annual GHG emissions over the COVID period (2019–2023). The pronounced drop in emissions in 2020 was due to the global lockdowns, which reduced CO_2 emissions from fossil fuel use, notably from global air transportation.

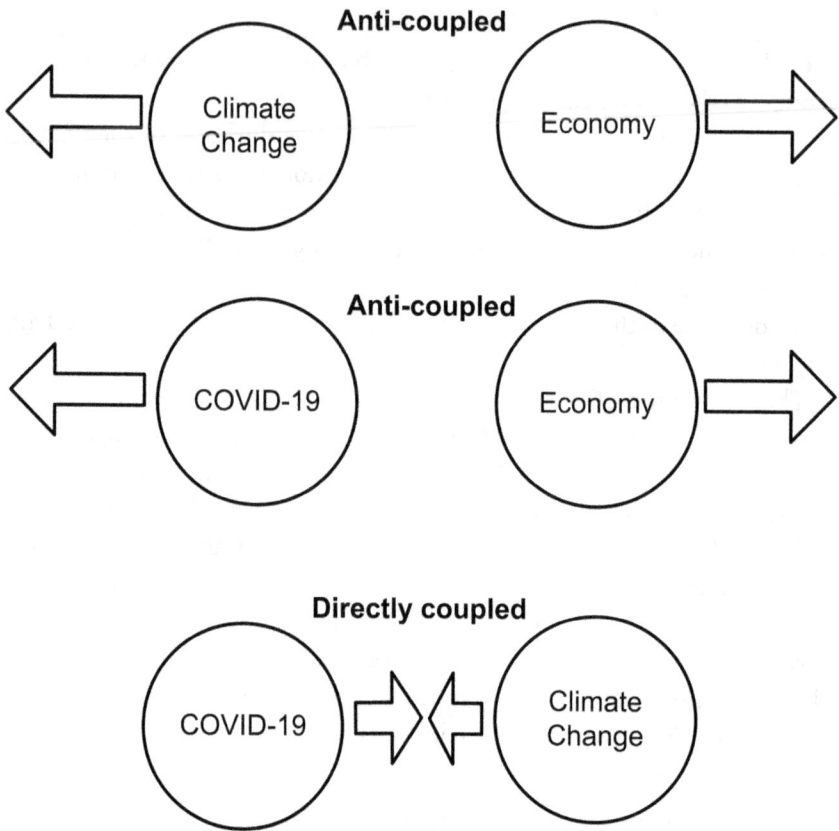

FIGURE 1.2 How the global crises are related to each other and the economy. They may be directly coupled or anti-coupled. (Figure credit: Asha Sinha.)

These countries had to weigh the objectives of the sanctions imposed on Russia against their energy needs. Most chose to maintain their trade links with Russia, although this clearly compromised the objectives of the sanctions. The economic consideration over-rode the sanctions issue.

1.4 THE IMPORTANCE OF PRIORITIES AND DECISION-MAKING

Whether an issue is given due attention depends not on how important it is, but what priority is accorded to it by the decision-makers at the time of consideration. Thus, climate change can be denied attention by the leaders of a country at a time when they deem that other issues have higher priority, even though climate change continues to pose an existential threat to human-kind.

As an illustrative example, consider the following case study (based on real-world events):

CASE STUDY: WHAT MATTERS MORE – ELECTRICITY FOR THE HOME OR CLIMATE CHANGE?

The residents of a certain suburb were up in arms about a series of power black-outs/brown-outs that had started to appear in their region. After making enquiries with the authorities, they learnt that they were caused by the ongoing phasing out of coal-fired power stations in the national grid that they were connected to.

The phasing out of the stations was part of the government's climate change strategy of replacing coal power with (lower emission) renewable power generation, and the reason for the blackouts was that the renewable energy replacements were not keeping up with the phasing out of coal power. This resulted in a net shortfall in the baseload power in the grid which resulted in blackouts/brownouts.

Even though most residents were sympathetic to the climate cause, they were adamant that the blackout issue was solved immediately, at any cost and by any means. To them, the reason for their demand was obvious. An assured power supply unquestionably had the highest priority over any other considerations, including climate change mitigation. In the event of a power black-out, the need for power became the most important consideration, and climate crisis faded into the background.

The government agreed with the residents and took emergency action by bringing back into operation enough gas-fired power stations to make up the shortfall. It did so even though it realised this was contrary to its own emissions reduction strategy. The next election was barely a year away, and it did not want to lose its voter support.

Note that the priority considered above is an economic priority, as adequate power supply is a basic economic need of the population.

This applies to other global crises as well, and the decision-makers at the national and international level are frequently faced with the task of assigning priorities to global crises relative to matters of other concern. Amongst democracies, it is invariably found that they give the economy the highest priority. And as the duration

This example can be generalised by saying that:

- Whether an action is taken or not will depend critically on its perceived priority at the time, and not on how important it is otherwise considered to be.
- It is very important to note that the priority is assigned by the decision-maker. In the case of climate change action, it is the government.

of an elected government is determined by its parliamentary term of office, it follows that the urgency for solving crises is always influenced by this term.

1.5 A (SEEMINGLY) INTRACTABLE PROBLEM

It was noted in Section 1.2 that COVID "ended" when its global emergency status was removed by WHO. In reality, it was still around. However, the world was effectively told by WHO to ignore it. Evidently, the hope was that the effects of the COVID virus would be subsumed within the other global viral infections such as influenza.

The above is an example of solving a problem at the level of human perception rather than the physical domain in which it exists. But as COVID still exists (the obvious reality test is the observation that people are still dying in hospitals due to the virus), one must find a justification for the seemingly arbitrary declaration by WHO.

Before attempting to do this, however, one must acknowledge that the pandemic had become an intractable global problem by the time of the declaration. The only justification could then be that the declaration was really a plea to the global community for a re-adjustment of human ethical standards to accommodate a trade-off between human health and economic needs at the global scale.

Intractable problems have not been uncommon within the disciplines of academia in the past, and their resolution invariably required complete (revolutionary) changes in the way scholars thought about problems. That is, they required conceptual revolutions, which involve changes to the conceptual frameworks of the relevant disciplines.

Physics was faced with several inexplicable phenomena during its development, which required revolutionary changes to the conceptual framework of the discipline for its resolution. Such milestones in the evolution of physics are provided by the resolution of the intractable problems of classical physics. The most notable of these were the ultra-violet (UV) catastrophe, the photo-electric effect, and the wave-particle duality, which were solved by the quantum revolution. Others were the need for a half-integral spin leading to the second revolution of physics, symmetry, the particle physics revolution and so on [9].

Take the example of the UV catastrophe. The observed spectrum of radiated power per unit wavelength of a black body follows a dome-shaped distribution. However, the equation derived using classical physics by Raleigh and Jeans gave an exponential curve that increased in intensity monotonically

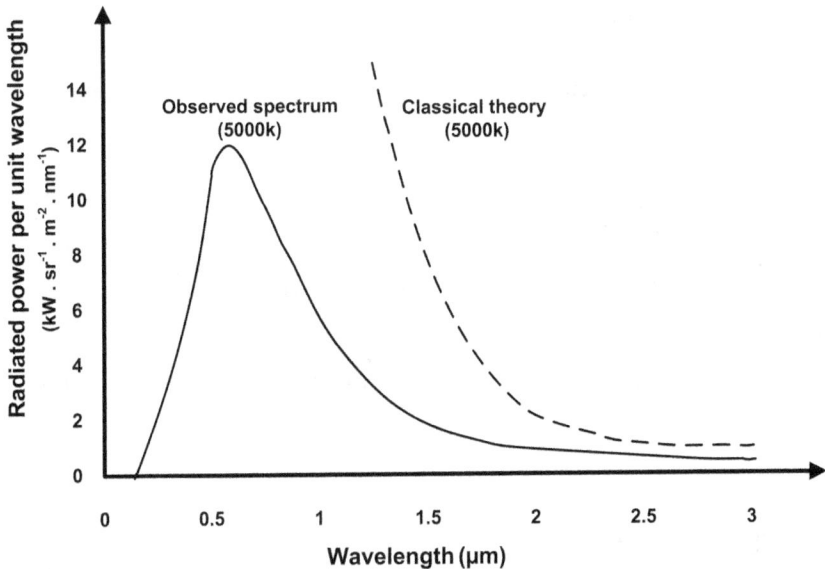

FIGURE 1.3 (After [10]) The ultra-violet catastrophe. The dome-shaped curve shows the true (i.e. observed) spectrum of the power radiated by a black-body at a temperature of 5000 K, while the exponential curve is what is predicted according to classical physics. It is clearly wrong as the area under the curve becomes infinite as the wavelength approaches zero. (Figure credit: Asha Sinha.)

as the wavelength approached zero. As the total power radiated by a black-body is equal to the area under the curve, this equation predicts that infinite power should be radiated by all black-bodies. This absurdly incorrect result of classical physics is known as the UV catastrophe. Plots of this (incorrect) classical equation and the observed (true) spectrum for a black body at a temperature of 5000 K are presented in Figure 1.3.

It took the introduction of the new revolutionary concept of energy quantisation by Max Planck to arrive at an equation that reproduced the correct (observed) black-body distribution. This laid the foundation stones for a completely new framework for physics, in the form of quantum physics, that solved the intractable problems of the black-body radiation together with others.

Will something similar be needed to discover the solution to the present climate crises? In Chapter 9, a possible conceptual framework approach to the climate crisis is suggested as worthy of consideration. But whatever the methodology chosen for the resolution of the climate issue, it must be a problem-solving approach that utilises new ways of framing the problem to yield possible solutions not evident before.

A primary aim of this book is to see what the world is doing to address the climate issue and to investigate whether there are better approaches to solving the problem. The discussions and arguments presented are grounded firmly in

- the understanding of the science and the laws of physics that determine the causes of climate change,

- analyses of the climate change mitigation actions taken at the national and global scales, and

- the socio-economic responses within countries and communities.

All this is underpinned by an awareness of the ethical standards that always guide the way decisions are made at the national and global scales.

1.6 SUMMARY

1. This introductory chapter begins with an overview of the global crises that are currently confronting the world and considers lessons we can learn from COVID-19, the current geopolitical scene as well as the climate crisis in an effort to find solutions for the latter.

2. COVID-19 was announced as an international public health concern in February 2022. By the time WHO declared an end to the emergency on 5 May 2023, it had recorded 765 million cases, killed 7 million people and wreaked havoc on the global economy. Amongst the lessons learnt were that modern economies depend critically on global supply chains, and that a crisis can be "ended" through a declaration by the relevant authority.

3. Two lessons learnt from the Russian invasion of Ukraine on 24 February 2022 and the election of Donald Trump to the US presidency in 2017 were that the economic interests of countries can be more important than the sovereign rights of nations, and that national politics can have a deciding role in the course of global climate action.

4. Global crises are frequently related to each other, and are always related to the global economy. Attempting to solve a crisis usually has an adverse effect on the economy (the two are anti-coupled). But the COVID-19 crisis had a beneficial effect on global warming, i.e. COVID and climate change were directly coupled to each other.

5. Whether an action is taken depends on its priority as perceived by the decision-maker at the time, and not on how important it is otherwise. Thus, climate change action may be ignored by decision-makers in favour of electricity supply to a community if there is a threat of power black-outs posed by the climate action.

6. The climate crisis is a (seemingly) intractable problem. Physicists have encountered several intractable problems in the history of the evolution of physics. These required revolutionary changes in the conceptual framework of physics to resolve. An example is the UV catastrophe predicted by classical physics in the spectrum of black-body radiation, which contributed to the development of quantum physics. Can similar changes be invoked in the resolution of the climate crisis? Whatever the case, it is certain that this pressing issue requires a problem-solving approach that is grounded firmly in science and a critical analysis of the outcomes of the climate change mitigation actions by the UNFCCC.

1.7 AN OVERVIEW OF THE REST OF THE BOOK

This chapter provided a picture of the global crises that are currently confronting the world, and showed how lessons could be learnt from them towards the resolution of the climate crisis.

Chapter 2 provides an overview of the current climate situation, including the status of global warming and the observed impacts of climate change. It also provides a first introduction to the impacts of climate change such as climate tipping points and ecosystem collapse, and their implications on food and bioenergy production. The need for new thinking to solve the (seemingly) intractable problem of the climate crisis is re-iterated.

Chapter 3 introduces the science of global warming, which is the cause of climate change. It explains global warming in terms of the Earth's Energy Budget and radiative forcing due to the emission of GHGs into the atmosphere. It further shows how the Earth's Energy Imbalance is leading to an increase in the Earth's Energy Inventory, and the resulting changes that are occurring in the Earth's climate system.

Chapter 4 provides an insight into the latest developments in the science behind climate change impacts. It begins with a brief introduction to the science of Extreme Event Attribution and the Forcing-Feedback model. The use of climate models to project the outcomes of climate impacts is described, and the chapter ends with an account of the nature of climate tipping points and the interactions between them.

Chapter 5 describes the history of the Paris Agreement and its outcomes in the conference of the parties (COPs). It begins by describing the history of the UNFCCC and its meetings, called the (COPs), which finally led to the Paris Agreement at COP21 in 2015. It goes on to outline the achievements of the COPs in climate change mitigation up to COP29.

Chapter 6 describes the strategies adopted for achieving net-zero emissions by 2050. The emissions reduction schemes, including the net-zero strategies, of the UK, Australia and Fiji, are provided as case studies. A case is made for a formal development process for net-zero strategies that includes the enumeration of their key requirements.

Chapter 7 shows how renewable energy provides a means of reducing net GHG emissions and outlines its role in the energy transition. It describes the range of renewable energy technologies that are available for the transition from fossil fuels to clean energy and provides case studies of the energy transition strategies of the UK and Australia. This is followed by overviews of proposed energy transition pathways by two independent global agencies. A critique is provided of nuclear energy as an alternative clean energy source for renewable energy.

Chapter 8 investigates the status of the Earth's climate around the warming temperature of 1.5°C in terms of global warming and several climate tipping points, with a special focus on the threats the impacts will pose to the survival and wellbeing of living organisms and ecosystems. It ends by speculating on the worst possible future climate scenario.

Chapter 9 proposes solutions to the climate crisis based on a new climate action framework that includes actions from other domains and provides a new basis for an all-inclusive climate action strategy.

REFERENCES

1. World Health Organization. COVID-19 public health emergency of international concern. 12 Feb 2022. Available from https://www.who.int/publications/m/item/covid-19-public-health-emergency-of-international-concern-(pheic)-global-research-and-innovation-forum. Accessed 7 Oct 2024.
2. UN. WHO Chief declares end to COVID-19 as a global health emergency. Available from https://news.un.org/en/story/2023/05/1136367. Accessed 16 Oct 2024.
3. United Nations. General assembly overwhelmingly adopts resolution demanding Russian Federation immediately end illegal use of force in

Ukraine, withdraw all troops. Available from https://press.un.org/en/2022/ga12407.doc.htm. Accessed 6 Oct. 2024.
4. CNN. The list of global sanctions on Russia for the war in Ukraine. Toh M. et al. 28 Feb 2022. Available from https://edition.cnn.com/2022/02/25/business/list-global-sanctions-russia-ukraine-war-intl-hnk/index.html. Accessed 6 Oct 2024.
5. Visual Capitalist. Which countries are buying Russian fossil fuels? N. Conte. 2 March 2023. Available from https://www.visualcapitalist.com/which-countries-are-buying-russian-fossil-fuels/. Accessed 8 Oct 2024.
6. US Department of State. Press statement by Michael Pompeo, Secretary of State. 4 Nov 2019. Available from https://2017-2021.state.gov/on-the-u-s-withdrawal-from-the-paris-agreement. Accessed 12 Sep 2024.
7. CNN. US begins formal withdrawal from the Paris Climate Accord. Drew Kann. 4 Nov, 2019. Available from https://edition.cnn.com/2019/11/04/politics/trump-formal-withdrawal-paris-climate-agreement/index.html. Accessed 12 Sep 2024.
8. WMO. Global temperature is likely to exceed 1.5 °C above pre-industrial level temporarily in next 5 years. 5 June 2024. Available from https://wmo.int/media/news/global-temperature-likely-exceed-15degc-above-pre-industrial-level-temporarily-next-5-years#:~:text=There%20is%20an%2080%20percent%20likelihood%20that%20the,new%20report%20from%20the%20World%20Meteorological%20Organization%20%28WMO%29. Accessed 14 Oct 2024.
9. Singh, A. 2021. Concepts and the Foundations of Physics. IOP Publishing. Available at https://pubs.aip.org/books/monograph/37/Concepts-and-the-Foundations-of-Physics. Accessed 11 Oct 2024.
10. Wikipedia. Black body radiation. Image source: https://en.wikipedia.org/wiki/Black-body_radiation#/media/File:Black_body.svg

CHAPTER 2

Climate Change

The Current Scenario

2.1 INTRODUCTION

As mentioned in Chapter 1, the world has been confronted with several global crises over the past few years. Amongst them, climate change is perhaps the most concerning because of the existential threat it poses to human-kind. So far, it has proved to be an intractable challenge.

A goal of this book is to examine alternative solutions that show more promise. Before embarking on the journey, however, one must take stock of the climate challenge as it exists today. An informed approach to its solution must begin with an examination of the scientific evidence available from the broadest perspective. This chapter begins the process by providing an overview of the climate crisis and the actions currently being taken to address it. It also provides a first introduction to key consequences of climate change, such as climate tipping points (CTPs) and economic impacts, that are treated in detail in later chapters.

The next section begins by highlighting the data on the current state of the crisis as reported by leading climate authorities and the success, to date, of the global action against climate change. The following two sections provide case studies of the latest extreme weather events as well as an overview of the success achieved in reducing fossil fuel emissions since COP26, including an evaluation of the impact of the US withdrawal from the Paris Agreement on 20 January 2025. The last three sections focus on

DOI: 10.1201/9781003531180-2

CTPs, as well as contemplating the nature of other poorly understood consequences. These include the economic impacts of climate change. The chapter ends with a call for new thinking to address the seemingly intractable climate change challenge.

Regular references are made throughout the chapter to more detailed treatments of the topics in later chapters.

2.2 WHAT THE AUTHORITIES SAY

The year 2024 was another year of broken climate records. It also saw continuing increase in GHG emissions despite all efforts to contain them. Some of the key climate metrics for the year, as recorded by leading climate authorities, are as follows:

KEY CLIMATE HIGHLIGHTS 2024

According to the Copernicus Climate Change Service (one of six thematic information services provided by the Copernicus Earth Observation Programme of the European Union) [1]:

- The 2024 average global surface temperature was a record 1.60°C above the pre-industrial (1850–1900) level, and the combined 2023/2024 average was 1.54°C.
- The last ten years have been the warmest ten years in recorded history.
- The sea surface temperature (over the extra-polar region of the world) was a record high of 20.87°C in 2024.
- The large number of extreme weather events recorded across the globe included heatwaves, floods, draughts and wildfires.
- The sea ice extent around Antarctica was the lowest on record over a large period of 2024.
- The atmospheric concentrations of carbon dioxide and methane were the highest in record at 422.1 ppm and 1897 ppb, respectively, and were 0.7% and 0.16% higher than the record concentrations reached in 2023
- During 2024, much of the globe experienced more days than average with at least "strong heat stress". On 10 July, around 44% of the globe was affected by "strong to extreme heat stress", which is 5% more than the average annual maximum.

Meanwhile

- The global annual GHG emissions have been rising steadily, reaching an all-time high of 57.1 Gigatons of CO_{2eq} in 2023 [2].

- But according to United Nations Environment Programme (UNEP) Emissions Gap Report 2024 Executive Summary [2]:
  - Progress in ambition and action since the initial NDCs has plateaued, and countries are still off-track to deliver on the globally insufficient mitigation pledges for 2030.
  - The emissions gap in 2030 and 2035 remains large compared with both pathways limiting warming to 1.5°C and 2°C. (For example, the emission gaps in 2030 for conditional NDCs are 11$GtCO_{2eq}$ and 19 $GtCO_{2eq}$ for the temperature limits of 2°C and 1.5°C, respectively).

These indicators of climate change and its impacts confirm that the climate crisis has been worsening rapidly over the last few years. Indeed, the situation has become so critical that the Earth's climate system is already showing early warning signals of several CTPs (irreversible changes in climate) that have been predicted by climate models (see Section 2.5 and Chapter 4).

It is important to note that the global warming temperatures mentioned above are not those referred to in the Paris Agreement. The average surface temperatures noted above are annual averages for the mentioned years only, whereas the limiting temperatures (of 2°C and 1.5°C) mentioned in the Paris Agreement are averages over 10 or 20 consecutive year periods (sometimes referred to as "decadal averages").

A key issue in determining average global warming using such extended periods is that taking (e.g.) a 20-year average to determine the year of exceedance of a particular temperature (e.g. the year in which the average temperature exceeded 1.5°C) means that this year can only be reported 10 years after the exceedance occurs.

The World Meteorological Organization (WMO), which produces annual State of the Climate Reports, has suggested three approaches to considering this issue in their 2024 Report that was updated for COP29 [3].

The average of these methods yields a temperature of 1.3°C for 2024 (see Chapter 5, Section 5.6.3 for more details). This is close to the value suggested by Copernicus, which is 1.36°C, and is expected to reach 1.5°C in June 2030.

2.3 THE CURRENT STATUS (AS IN JANUARY 2025)

Two major events impacting climate change and climate action in 2024 and early 2025 were the Los Angeles (LA) wildfires and President Trump's withdrawal of the US from the Paris Agreement.

2.3.1 LA Wildfires

A chronology of the LA fires, which started in the Palisades and Eaton on 7 January 2025 and were finally contained on 27 January, is given by Meredith Delio and Julian Kim, of the ABC news [4]. By the time the fires were contained, they had left 16,200 structures damaged or destroyed and 29 people dead over 45 square miles (116.6 square kilometres) of densely populated LA County. The scene of devastation left behind has been described by some as "apocalyptic". The causes of these wildfires have been listed as drought conditions, strong (Santa Anna) winds, human activity and climate change [5].

California has experienced a decade-long drought in the recent past. This was followed by an unusually wet period between 2022 and 2023, which induced rapid growth of vegetation. This dried out later to provide copious amounts of fuel for wildfire conditions. This "climate whiplash" is cited as the primary cause of the LA fires [6]. A recent review of scientific studies of the phenomenon has been carried out by Swain et al. [7].

Whether an extreme climate event is caused by climate change can be assessed by a recently developed statistical technique known as *Extreme Event Attribution* (*EEA*). This method was first used by Stott, Stone and Allen [8] to associate the European heatwave of 2003 with human-induced climate change. Since then, hundreds of groups have been actively engaged in attributing extreme weather events to climate change. To date, no similar analyses have emerged of the LA fires in the literature. A brief introduction to the science of extreme weather attribution is provided in Chapter 4.

2.3.2 US Withdrawal from the Paris Agreement

Global climate action was dealt a severe blow when, on the 20th of January 2025, the newly inaugurated President Trump signed an executive order withdrawing the US from the Paris Agreement [9].

Titled "Putting America first in International Environmental Agreements", the document ordered the immediate withdrawal of the United States from the Paris Agreement under the United Nations Framework Convention on Climate Change.

In addition, the order

- ceased or revoked "any purported financial commitment made by the United States under the United Nations Framework Convention on Climate Change", and

- revoked and rescinded the US International Climate Finance Plan.

It is immediately clear that one of the most serious repercussions of this executive order will be on the financial support received by the developing Parties of the Paris Agreement for their mitigation and adaptation efforts. This is because the developed Parties of the Agreement (which includes the US) are obliged under the Agreement to provide such support to developing Parties (see Articles 4, 9 of the Paris Agreement – Chapter 5).

2.4 STATUS OF CLIMATE ACTION

The first concerted action to reduce emissions from fossil fuels, which are the largest emitters, was taken in COP26, resulting in a commitment to phase down coal power.

However, no further progress has been made on the initiative since. This is despite the repeated warnings of the widening emissions gap by UNEP, which stated in its Emissions Gap Report 2024 [2] that to limit global warming to 1.5°C, the annual emissions in 2030 would have to be 22 $GtCO_{2eq}$ lower than what the current policies predict. It is most concerning, therefore, to note that at the last COP in Baku (COP29), the issue of transitioning away from fossil fuels was indeed prevented from entering the meeting agenda (see Chapter 5, Section 5.7.1 for more details).

With the election of Donald Trump as the new US president, prospects of reducing fossil fuel emissions have worsened. Amongst the nearly hundred executive orders that President Trump issued on 20 January was one to re-commence drilling of the US fossil fuels reserves [10]. The order declared a state of national energy emergency and ordered the

> "the identification, leasing, siting, production, transportation, refining, and generation of domestic energy resources, including, but not limited to, on Federal lands."

where "energy" is defined as fossil fuels, uranium, geothermal, hydro and critical minerals.

In addition, it required the Environmental Protection Agency to issue emergency fuel waivers to allow the year-round sale of E15 gasoline in the event of any projected shortfalls in gasoline supply in the country.

The impact of these decisions is likely to be far-reaching on the global scale. The US is second amongst the world's three largest emitters (China, the US and India), which are together responsible for around 40% of the total global emissions [11]. Given that President Biden had ambitious goals for US emissions reductions [12], the combined effect of the US withdrawal from the Paris

Agreement and President Trump's declaration of the energy emergency is likely to deliver a devastating blow to the global emissions reduction efforts.

2.5 CONSEQUENCES OF CLIMATE CHANGE

The previous sections highlighted some of the severe impacts of climate change that are currently being experienced globally. They included the extent of land and sea surface warming and the related droughts, heat waves, wildfires and sea ice sheet extent. The subsections that follow focus on two additional key consequences of climate change, and go further to speculate on the nature of the (yet unforeseen) socio-economic consequences of a changing climate system.

2.5.1 Climate Tipping Points (CTPs)

In addition to the extreme weather events noted in Sections 2.2 and 2.3, the impacts of climate change include CTPs, which herald irreversible changes to the Earth's Climate System at certain levels of warming. At these temperature thresholds, certain parts of the Earth's climate system (called *tipping elements*) undergo irreversible changes.

An example is the West Antarctic sea-ice sheet, which is part of the Earth's *cryosphere*. This sheet undergoes annual cycles of expansion during the winter months and contraction during the summer. When global warming reaches a certain threshold temperature (which is close to 1.5°C in this case), climate modelling predicts the irreversible disappearance (or collapse) of the sheet.

Further examples of well-documented CTPs are given in Table 4.4 of Chapter 4. McKay et al. [13] have identified 16 such CTPs. Six of these are likely within the 1.5°C–2°C temperature range.

CTPs pose the greatest threat to the future of the Earth's climate system, and will be discussed at length in later chapters.

2.5.2 Ecosystem Collapse

Several of the CTPs predicted by climate models and monitored by Earth observation systems involve tipping elements related to the Earth's biosphere. As this provides the habitats and accommodates the ecosystems that sustain all life forms, breaching the temperature thresholds of these tipping points may lead to the collapse of such ecosystems. As all ecosystem-related tipping points known to date lie below the 2°C temperature limit, such CTPs require the most urgent attention.

Two examples of ecosystem collapse are *Coral Reef Bleaching* events [14] and *Forest Ecosystem Collapse* [15]. In the first, the rising ocean surface temperatures result in the death of coral reefs and the consequent loss in marine biodiversity. Forest ecosystem collapse results from unusually high land surface air temperatures and drought conditions, leading to the collapse of forest ecosystems and endangering ecosystems that support land-based human and animal food chains.

Ecosystem collapse will be considered again in Chapters 4, 8 and 9.

2.5.3 Unforeseen Consequences

The climate impacts considered above are all physical effects, i.e. changes in the climate as measured by physical instruments. But apart from the purely physical effects, climate change can have consequences on other aspects of human wellbeing as well.

As mentioned in Chapter 1, global crises are related. It should not be a surprise therefore if the global climate action is impacted by political crises such as changes in the established geo-political order. It was also noted that economics seemingly plays a central role in all crises of national and global proportions.

More specifically, it can be asserted that national crises always impact economies, and decisions for addressing national crises are invariably determined/influenced by economic priorities. To obtain a better understanding of the solution to the climate crisis, it is therefore important to have a clear perspective of the economic ramifications of climate change.

A similar consideration should be accorded to changes in the geo-political landscape. Revelations of the US-centric policies of the new US presidency are clear signs of an emerging shift in the global geo-political order. The initial impacts of this shift have been noted above. But as this new geo-political landscape is still evolving, it is difficult to model its continuing influence on climate action. It is prudent therefore to restrict our attention to what we can analyse with reasonable certainty. One such example is the economic consequence of climate change in a world beyond 1.5°C. So what are the possible economic impacts of the new climate regime after the 1.5°C threshold?

The obvious effects are supply chain disruptions. A prelude of such economic impacts of global crises was provided by the shock that COVID delivered to the global economic system early this decade. But one can also envisage other possibilities which are too subtle to comprehend fully at these early stages. One such effect is a widening of the divide between rich and poor nations, perhaps leading to behavioural changes amongst both people and their communities. Another has to do with the limited availability of resources for surviving the (highly likely) extreme weather conditions of the new climate equilibrium.

2.6 IMPLICATIONS ON FOOD AND BIOENERGY PRODUCTION

One of the most direct impacts of the ecosystem collapse referred to in Section 2.5.2 is likely to be on agricultural production, leading to consequences on human food chains. There will be similar impacts on the ecological food chains within the biosphere. The significance of such ecosystem collapse cannot be under-estimated, as they have the potential of leading to whole of biosphere instabilities. Such possibilities are the subject of active modelling using Earth Systems Models (ESMs) [16].

The possibility of biosphere collapse is real and poses an existential threat that cannot be ignored. Arguments to support this thesis will be presented in the last two chapters.

2.7 NEED FOR NEW THINKING

Section 2.4 reveals that the current climate action is failing to address the urgent need to find ways of dealing with the worsening state of the global climate. The situation has been exacerbated by the US withdrawal from the Agreement and its decision to commence mining its fossil fuel reserves to make additional energy available for its internal use.

It is clear that current methods of addressing the climate challenge are not working. There is a need for new thinking to deal with what has seemingly become an intractable challenge. This requires a fresh start and a new way of framing the problem that widens the scope and provides new insights into possible new solutions. This will be the subject of the final two chapters.

2.8 SUMMARY

1. According to the Copernicus Climate Change Service (an information service provided by the EU), in 2024 the Earth's global average surface temperature had reached a record 1.6°C above pre-industrial levels, sea surface temperatures had peaked at a record high of 20.87°C, the sea ice extent around Antarctica had reached a record minimum, the atmospheric carbon dioxide and methane contents were the highest on record and the extent of strong heat stress felt by humans globally had also reached a record high.

2. The LA wildfires that started on 7 January 2025 damaged or destroyed 16200 buildings and structures, killing 29 people over 116 square kilometres of the LA County in California, USA.

3. Climate action was dealt a severe blow on 20 January 2025 when the US withdrew from the Paris Agreement. The withdrawal was followed by a declaration of a state of national energy emergency by re-elected President Trump.

4. Amongst the consequences of climate change are CTPs, which are temperature thresholds at which a change in the Earth's climate becomes irreversible. Six of these CTPs are likely to occur within a warming temperature close to 1.5°C.

5. Amongst the CTPs are ecosystem-related tipping points, sometimes called *ecosystem collapse*.

6. Amongst the unforeseen consequences of climate change are their specific impacts on the global economy. Impacts such as supply chain disruptions are obvious. However, others remain speculative.

7. Ecosystem collapse becomes possible below a 2°C warming limit and will impact human and animal food production. The possibility that such events may lead to total biosphere collapse is not unreal.

REFERENCES

1. Copernicus. The 2024 annual climate summary. Global Climate Highlights 2024. 10 Jan 2025. Available from https://climate.copernicus.eu/global-climate-highlights-2024. Accessed 3 Mar 2025.
2. UNEP Emissions Gap Report 2024. No more hot air ... please! Available from https://www.unep.org/resources/emissions-gap-report-2024. Accessed 3 Mar 2025.
3. WMO. State of the climate 2024 – Update for COP29. Available from https://wmo.int/publication-series/state-of-climate-2024-update-cop29. Accessed 3 Jan 2025.
4. ABC News. Los Angeles wildfires timeline: How the deadly blazes unfolded. Meredith Deliso and Julian Kim. 28 Jan 2025. Available from https://abcnews.go.com/US/los-angeles-wildfires-timeline-palisades-eaton/story?id=117643473. Accessed 16 Feb 2025.
5. Internet Geography. Wildfires in California: Causes, effects and responses. 13 Jan 2025. Available from https://www.internetgeography.net/wildfires-in-california-causes-effects-and-responses/. Accessed 16 Feb 2025.
6. BBC. Climate 'whiplash' linked to raging LA fires. Matt McGrath. 10 Jan 2025. Available from https://www.bbc.com/news/articles/c0ewe4p9128o. Accessed 16 Feb 2025.

7. Swain, D. L. et al. Hydroclimate volatility on a warming Earth. Nature Reviews Earth and Environment 6 (2025) 35–50. Available from https://www.nature.com/articles/s43017-024-00624-z.epdf?sharing_token=jj_XNSm-QH1g15lA6CB1_tRgN0jAjWel9jnR3ZoTv0O9RZ3Zpesp9Svwudh0S7m0ji A4yzuG7jBaggtdEjGWPZrA3_NRiYjrqQgmpIyODy8XLQ93lqs65bQek9-QhceIvet0w4ND0s65f-s0kDK9DBjBLYaoOATK3vqu8juO2eWOF gQhfBLabslQxdSLmBguzseRYeuQbOKR_JsisJ-hRsmv4vi0aVQv-zDXWA8mEO7C_E9tiBxJiD6T9F2zVEuG&tracking_referrer=www.bbc.com. Accessed 16 Feb 2025.
8. Stott, P., Stone, D. & Allen, M. Human contribution to the European heatwave of 2003. Nature 432 (2004) 610–614.
9. The White House. Putting America first in international environmental agreements. 20 Jan 2025. Available from https://www.whitehouse.gov/presidential-actions/2025/01/putting-america-first-in-international-environmental-agreements/. Accessed 16 Feb 2025.
10. The White House. Declaring a national energy emergency. Executive order. 20 Jan 2025. Available from https://www.whitehouse.gov/presidential-actions/2025/01/declaring-a-national-energy-emergency/. Accessed 27 Feb 2025
11. World Resources Institute. This interactive chart shows changes in the world's top 10 emitters. March 2023. Available from https://www.wri.org/insights/interactive-chart-shows-changes-worlds-top-10-emitters. Accessed 20 Feb 2025.
12. Associated Press. Biden … sets ambitious climate goal. 20 Dec 2024. Available from https://apnews.com/article/biden-climate-trump-paris-warming-un-e7c6af28dc013460a3aa66a644c55145. Accessed 20 Feb 2025.
13. Armstrong McKay, David, & Staal, Arie et al. Exceeding 1.5 C global warming could trigger multiple climate tipping points. Science 377 (6611) (9 Sept 2022). DOI: 10.1126/science.abn7950.
14. NOAA. 4th global coral bleaching event. 15 April 2024. Available from https://www.noaa.gov/news-release/noaa-confirms-4th-global-coral-bleaching-event. Accessed 11 Feb 2025
15. ABC News. Fears of another 'forest collapse' event in Western Australia after record dry spell Stateline. Briana Shepherd. 11 April 2024. Available from https://www.abc.net.au/news/2024-04-11/ecologists-warn-potential-forest-collapse-event-wa/103682304. Accessed 11 Feb 2025.
16. UKESM. What is the Earth System and how do you model it? Available from https://ukesm.ac.uk/what-is-the-earth-system-and-how-do-you-model-it/. Accessed 4 Mar 2025.

CHAPTER 3

The Cause of Climate Change

3.1 INTRODUCTION

As indicated in Chapter 1, a primary aim of this book is to scrutinise and discuss in detail the methodology the global community has adopted to solve the climate problem. This adopted recipe is clearly specified in the Paris Agreement, which asserts that the issue of global warming is to be solved by reducing net global emissions to zero by the year 2050.

In itself, the Agreement is merely a policy document. Its ultimate success lies in its implementation, and to succeed, one must adopt a methodology that works. This requires a *problem-solving approach*. One such approach is to begin by identifying the *cause* of the problem. Once this has been determined, the next step would be to ascertain what the *requirements for a viable solution* are.

Several problem-solving techniques exist that can be easily adapted to addressing the climate issue. One such method is a *systems-analytical approach*, which starts with a requirements analysis and continues the process by designing a *logical solution* first and using it to construct a *physical design* which acts as the blueprint for the final implementation. The implementation of the solution to a problem is frequently carried out according to a *roadmap*. In the present case, the roadmap translates to the *net-zero strategy* for achieving a successful solution to the climate issue.

This book is an investigation into how the global community is implementing its chosen roadmap to climate success. Such a roadmap

 DOI: 10.1201/9781003531180-3

must begin by identifying the cause of the problem, which is the subject of this chapter.

Climate change is caused by global warming, which occurs when the earth's energy budget is disturbed. The next section thus begins by introducing the science of global warming. It does so by presenting a simple model of how the *earth's energy budget* works. The following section shows how measurements of the *Earth's Energy Imbalance* (EEI) by the NASA-sponsored CERES project have led to a better understanding that provides a necessary correction to the simple model.

Greenhouse gases (GHGs) are the agents that cause the "disturbance" to the energy budget by changing the radiative flux designated as OLR (see later for definition). It is this *radiative forcing* that leads to global warming. The final sections of the chapter show how GHGs produce radiative forcing and elaborate on further properties of the GHGs, including how their warming effects can be compared and followed over the years.

3.2 HOW CLIMATE CHANGE OCCURS

3.2.1 The Earth's Energy Budget

Global warming occurs when the earth's energy budget is upset. To understand how this happens, we can begin by first noting that the earth receives almost all its energy from the sun and then using a simple model of the energy budget that can be refined later.

To summarise what happens during global warming, we begin by noting that the earth warms up as it absorbs part of the energy it receives from the sun. The earth in turn re-radiates energy outwards, and energy balance is reached when the net incoming (short wavelength) radiation from the sun becomes equal to the net outgoing (long wavelength, or infra-red) radiation from the earth. This simple model assumes that nothing else happens to the earth during this energy exchange apart from the warming up of the earth.

The addition of GHGs to the atmosphere reduces the *Outgoing Long wavelength Radiation (OLR)* from the earth, causing the earth's temperature to rise (i.e. produces global warming) till the balance is re-established. The situation may be visualised with the help of Figure 3.1, which depicts the sun above emitting solar radiation (light grey arrows) and the earth-atmosphere system emitting infra-red radiation (dark grey arrows).

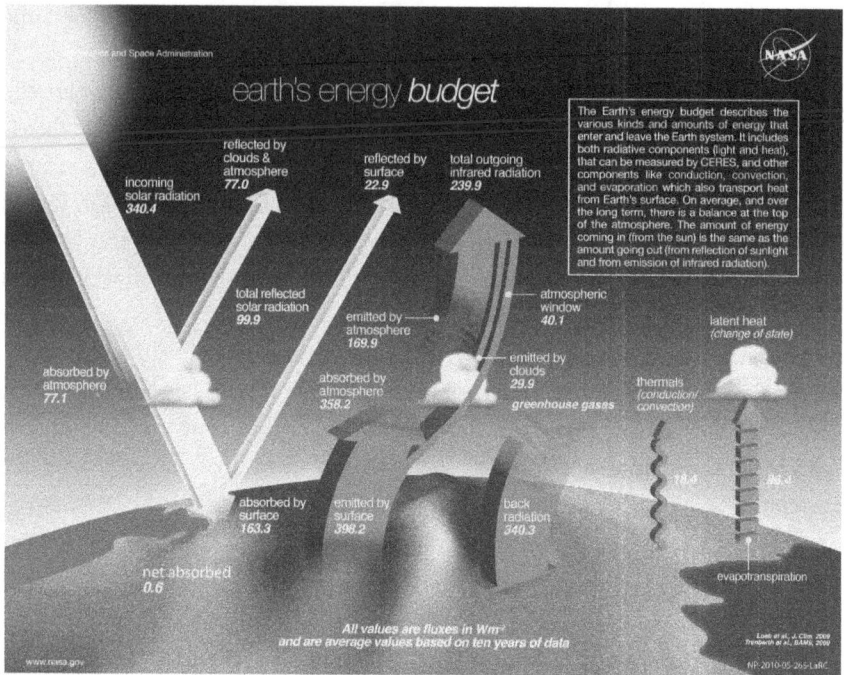

FIGURE 3.1 Schematic diagram from NASA of the earth's energy budget, showing incoming short-wavelength solar radiation in light grey and outgoing infrared radiation (dark grey) from the earth. This diagram can be used to visualise how the simple model of the earth's energy budget works if we assume that no other change occurs to the earth apart from the warming. (Picture source: [1].)

To obtain a fuller explanation of the energy budget, we first need to have a closer look at the terminology used to describe it. In particular, we need to know

- what is meant by "radiation"

- what is black body radiation

- what is meant by short and long wavelength.

The energy balance is best envisaged in terms of the sun's *radiative flux*. This is the value of its emitted power per unit surface area perpendicular to the direction of radiation. At the top of the earth's atmosphere (TOA), which is approximately 1.5×10^6 km away from the sun, the sun's radiative flux (called the solar constant) is 1360 W/m². This flux is intercepted by the earth's cross-sectional area. However, over a complete day, it actually falls on the whole spherical surface of the earth, which has four times the area of the earth's cross-section. Therefore, the mean effective flux becomes a fourth of solar constant, or 340 W m^{-2}.

Both the sun and the earth behave as black-body radiators, which have the property that the total radiant power they emit is given by Stefan's Law, which is

$$P = \sigma A\, T_S^4 \qquad (3.1)$$

where P is the total power emitted by the surface, $\sigma = 5.670 \times 10^{-8}$ W/m^{2}K^4 is the Stefan-Boltzmann constant, A is the surface area and T is the temperature in Kelvins.

The sun's (and the earth's) radiative flux is thus proportional to the fourth power of their surface temperatures. Expression 3.1 shows that all black-bodies give out radiation, no matter how small their temperature.

To obtain greater insight into the nature of the radiation from the sun and the earth, note that each gives out radiation that is spread over a spectrum of energies, which can be characterised either by their frequencies or wavelengths. A property of such black-body spectra is that the peak wavelength of a spectrum is inversely proportional to the absolute temperature T of the body. This is known as Wein's Law, which states that

$$T\,\lambda_{max} = \text{constant}\,(= 0.2898\,\text{cm-K}) \qquad (3.2)$$

where λ_{max} is the wavelength at which the energy spectrum peaks and T is the temperature in Kelvins.

As the sun's surface temperature is about 5500°C, the peak wavelength of light it emits is far shorter than that of the peak radiation of the earth, which has a mean surface temperature closer to 15°C. The radiation from the sun is thus termed *short wavelength* (and occurs mostly in the visible), while the black-body radiation emitted by the earth is called *long wavelength radiation* (and occurs in the *infra-red*). It is also sometimes called *thermal radiation*.

We can now provide a fuller explanation of global warming according to the simple model and with the aid of Figure 3.1 as follows:

At TOA, 340 Wm^{-2} short wavelength solar radiation enters the atmosphere. Some is reflected back by the clouds and the atmosphere, and some when it reaches the earth's surface. The rest (called *Absorbed Solar*

Radiation (ASR)) is absorbed by the earth's surface, causing its temperature to rise.

The warmer earth now produces outgoing long wavelength radiation (OLR) in the infra-red. Some of this is absorbed by the atmosphere and its contents, while some escapes directly into outer space. There is also some energy (in the form of heat) transferred from the earth's surface to the atmosphere through the processes of conduction, convection and moisture loss.

Meanwhile, the contents of the atmosphere (including the clouds and all their aerosol contents) also emit radiation (in all directions) through the same process of black body radiation as the sun and earth do. Because they are also at low temperature, they emit in the long wavelength (infra-red) region of the spectrum.

At the top of the atmosphere (TOA), the net result of all these processes, according to the simple model of the earth's energy budget, is:

$$ASR = OLR \tag{3.3}$$

where both ASR and OLR are determined at the TOA.

Global warming occurs when addition of human-produced GHGs to the atmosphere upsets the balance by reducing the value of OLR. The earth's surface has to warm up to re-establish the equilibrium at a higher temperature $T + \Delta T_s$, where ΔT_s is the global warming temperature.

The main points of the simple model of global warming are repeated in Box 3.1.

BOX 3.1   HOW THE SIMPLE MODEL EXPLAINS GLOBAL WARMING

- The sun and earth both behave as black-body radiators, which means that the power they radiate increases rapidly with their temperature.
- The peak wavelengths of such black-body spectra vary inversely with the surface temperatures of the bodies. This means that hotter bodies have shorter peak wavelengths. The cooler earth thus gives out longer wavelength radiation (in the infra-red).
- At the top of the atmosphere (TOA), 340 W/m² of radiative flux arrives from the sun.
- Part of this is reflected back by the clouds and the atmosphere, and some more is reflected back when the sunlight reaches the earth's surface.
- The remaining solar radiation is absorbed by the earth's surface and is called the Absorbed Solar Radiation (ASR). This causes the earth's surface temperature to rise.

- The warming earth produces Outgoing Long wavelength Radiation (OLR). Some of this is absorbed by the atmosphere and its content, while the rest escapes into outer space.
- The contents of the atmosphere, which have temperatures close to that of the earth's surface, emit long-wavelength radiation (in all directions)
- According to the simple model of the earth's energy budget, at the TOA, the net result of all these processes is that the OLR becomes equal to the ASR at equilibrium.
- If some GHG is now added to the atmosphere, it absorbs some of the OLR travelling upwards from the surface, thus reducing its value. As the ASR has remained constant, the energy balance is upset, and the earth is absorbing more radiation than it is emitting.
- To re-establish the balance, the earth heats up by the temperature ΔT, which is called the global warming temperature.

3.2.2 The Earth Energy Imbalance (EEI)

Measurements made by the Clouds and the Earth's Radiant Energy System (CERES) project administered by NASA [2] show that Expression 3.3 is in fact never quite true. This would have been correct if the assumption made by the simple model in Section 3.2.1 were true, i.e. if there were no other way in which energy was being added or subtracted from the budget.

The CERES experiment is a mainly satellite-based project that measures the solar-reflected and earth-emitted radiation from the atmosphere and the earth's surface at the TOA. Its main aim is to understand the role that clouds play in climate change.

The project, which began in the 1970s, has revealed that the values of ASR and OLR at the TOA are not exactly equal. This can be accounted for by saying that a small fraction of the ASR is being absorbed continuously by certain parts of the Earth. More specifically, the elements of the *Earth's climate system*, consisting of the oceans (called the hydrosphere), the land surface (part of the lithosphere), surface ice content (the cryosphere) and the atmosphere, are absorbing energy in the form of thermal energy. This additional energy flow needs to be accounted for.

This is done by defining an additional energy term called the *Earth's Energy Imbalance (EEI)* as the difference between ASR and OLR and saying that EEI adds to the Earth's energy inventory and contributes to the impacts of climate change (See [3] for further information).

A better equation for the earth's energy budget therefore is:

$$ASR = OLR + EEI \tag{3.4}$$

So where does the missing energy go? As the earth's energy budget is simply another expression of the law of conservation of energy, the "missing" energy must be appearing as some other form of energy. In fact, the quantity labelled as EEI is simply the thermal energy, noted above, that is being added to the earth's climate system inventory. The simple model had not accounted for this and was thus not quite correct!

According to the IPCC Report [3], the global climate system energy inventory increased by 282 ZJ in the 1971–2006 period and by 152 ZJ in the 2006–2018 period, corresponding to 0.5 and 0.79 W m⁻² radiative imbalance, respectively. This increase is expected to continue till the end of the 21st century (Note: 1 ZJ = 10^{21} J).

3.2.3 Greenhouse Gases as Sources of Radiative Forcing

As the simple model evoked above shows, global warming occurs whenever GHGs are added to the atmosphere. These gases are able to absorb some of the OLR emitted from the surface, thus reducing the net OLR reaching the TOA.

We can re-state the above by saying that the warming was "forced" by the GHG through reducing the outward flux of energy from the earth. The reduction in the flux is called the *radiative forcing* (ΔF) due to the GHG. Because the change it produces is a radiative flux, it is measured in Watts/m².

3.2.4 Radiative Forcing and Global Warming

The National Oceanic and Atmospheric Administration (NOAA) gives expressions that relate the concentration of a GHG to its radiative forcing ΔF. In the case of carbon dioxide, this is given by [4]

$$\Delta F = \alpha \ln(C / C_o) \tag{3.5}$$

where C_o is the pre-industrial GHG concentration, C is the present GHG concentration and α is a polynomial in the difference between these concentrations.

The global warming ΔT_s can then be related to the radiative forcing ΔF by the expression

$$\Delta Ts = \lambda \Delta F \qquad (3.6)$$

where λ is called the climate sensitivity parameter.

3.2.5 Global Warming Potentials

Not all GHGs produce the same global warming. Thus, a tonne of one GHG will generally produce a different amount of global warming as compared to another. The ability of a GHG to produce global warming is characterised by its *global warming potential (GWP)*. This is defined as the amount of radiation one tonne of the gas will absorb over either 100 or 20 years of its existing life. These periods are called the "time horizon" of a GHG, with the 100-year time horizon being quoted most commonly.

The GWPs of individual GHGs are usually expressed by comparing them with the GWP of carbon dioxide, which is assigned a value of unity. Table 3.1 lists the 100-year GWPs of the most important GHGs as noted in the IPCC Fifth Assessment Report (AR5).

3.2.6 Annual Greenhouse Gas Index (AGGI)

There are several GHGs present in the atmosphere at any one time, and global warming is produced through the radiative forcings due to all of these. It is therefore natural to define a quantity that indicates the total global warming due to the atmospheric concentrations of all GHGs. This

TABLE 3.1   Some Representative GHGs and Their 100-yr GWPs

Name	Formula	GWP (100-yr Time Horizon)
Carbon dioxide	CO_2	1
Methane	CH_4	28
Nitrous oxide	N_2O	265
CFC-12	CCl_2F_2	10,200
CFC-114	$CClF_2CClF_2$	8590
Carbon tetrachloride	CCl_4	1730
Halon 1211	$CBrClF_2$	1750
HFC23	CHF_3	12,400
HCF125	CHF_2CF_3	3170
HFC143a	CH_3CF_3	4800

Source: [5].

TABLE 3.2 Radiative Forcing, CO_{2eq} Concentration and Annual Greenhouse Gas Index (AGGI) of Atmospheric GHGs for Selected Years. (W/m² = Watts per square meter; ppm = parts per million)

Year	\multicolumn{6}{c}{Radiative Forcings for GHGs (W m⁻²)}						Total Forcing	CO_{2-eq} (ppm)	AGGI
	CO_2	CH_4	N_2O	CFCs	HCFCs	HFCs			
1979	1.025	0.504	0.088	0.175	0.008	0.001	1.798	388	0.787
1980	1.058	0.509	0.088	0.185	0.009	0.001	1.850	392	0.810
1990	1.240	0.564	0.112	0.296	0.020	0.003	2.285	425	1.00
2000	1.511	0.591	0.133	0.316	0.035	0.008	2.593	450	1.135
2010	1.791	0.602	0.156	0.299	0.051	0.023	2.921	478	1.278
2015	1.939	0.617	0.171	0.289	0.058	0.032	3.107	495	1.359
2020	2.110	0.636	0.185	0.279	0.061	0.044	3.316	515	1.431
2021	2.140	0.643	0.189	0.276	0.061	0.046	3.356	519	1.469
2022	2.170	0.650	0.193	0.274	0.061	0.049	3.398	523	1.487

Source: [4].

provides a quantitative indicator of the global warming and facilitates its tracking over the years.

Such a measure is the NOAA *Annual Greenhouse Gas Index (AGGI)*. It relates the sum of the radiative forcings due to all the atmospheric GHGs and tracks how atmospheric GHG concentrations have been changing over the years as compared to the year 1990 by assigning an AGGI index of unity to this year.

Table 3.2 shows the radiative forcings, CO_{2-eq} concentration and AGGI for selected years from 1979 to 2022.

Table 3.2 reveals that in 2022, global GHG concentrations had reached 523 ppm, and global warming-inducing radiative forcing had increased by 49% over its value in 1990.

3.3 SUMMARY

1. A key aim of this book is to investigate the methodology adopted by the global community to address the climate crisis. This methodology takes the form of the net-zero strategies of the nations of the world. Such strategies are best formulated using problem-solving approaches, which begin by examining the cause of the problem and ascertaining the requirements for viable solutions.

2. This chapter begins by examining global warming, which is the cause of climate change.

3. The earth's energy budget is the balance between ASR and the net OLR from the earth. At equilibrium, these two quantities are equal.

4. According to a simple model, the addition of GHGs to the atmosphere reduces the OLR, which causes the earth to warm up and emit more radiation till the balance between ASR and OLR is re-established.

5. Measurements of the radiation at the top of the atmosphere by NASA over the last few decades reveal that the values of ASR and OLR have never been quite equal over this period. The difference is called the EEI and is explained by noting that the energy inventory of the earth's climate system (consisting of the land, sea, ice and atmosphere, all of which are strongly coupled to the biosphere) has been increasing by a similar amount over these years.

6. The process by which the GHGs disturb the earth's energy budget is known as radiative forcing (ΔF). This is related to the atmospheric concentration of the GHGs. Global warming (ΔT) is related to ΔF through a simple relation.

7. The same amounts of GHGs will, in general, produce different amounts of global warming. The ability of a GHG to produce global warming is characterised by its *GWP*. This is defined as the amount of radiation one tonne of the GHG will absorb in either 100 or 20 years of its lifetime.

8. Global warming is due to the sum of the global warming produced by each GHG present in the atmosphere. A quantity that measures this total warming in any one year is the *AGGI*. This is given a value of one for the carbon dioxide present in the atmosphere in 1990. The AGGI is a convenient tool to follow how global warming has changed over the years.

REFERENCES

1. Wikipedia. Earth's energy budget. Available from https://en.wikipedia.org/wiki/Earth%27s_energy_budget#/media/File:The-NASA-Earth's-Energy-Budget-Poster-Radiant-Energy-System-satellite-infrared-radiation-fluxes.jpg. Accessed 7 Aug 2024.
2. NASA CERES. What is CERES? Available from https://ceres.larc.nasa.gov/. Accessed 7 Aug 2024.

3. Forster, P. et al. 2021. The Physical Science Basis. Contribution of Working Group I to the Sixth Assessment Report of the Intergovernmental Panel on Climate Change (Masson-Delmotte, V. et al. (eds.)). Cambridge University Press, Cambridge, United Kingdom and New York: 923–105. Available from https://www.ipcc.ch/report/ar6/wg1/downloads/report/IPCC_AR6_WGI_Chapter07.pdf

4. Global Monitoring Laboratory. NOAA annual greenhouse gas index (AGGI). https://gml.noaa.gov/aggi/aggi.html. Accessed 16 Aug 2024.

5. Global Warming Potential Values. Greenhouse gas protocol. https://ghgprotocol.org/sites/default/files/ghgp/Global-Warming-Potential-Values%20%28Feb%2016%202016%29_1.pdf. Accessed 19 Feb 2024.

CHAPTER 4

The Impacts of Climate Change

4.1 INTRODUCTION

As noted in Chapter 2, the world is experiencing a dramatic acceleration in the nature, number and intensity of extreme weather and other climate events caused by climate change. Our aim is to know why this is happening and look for ways of mitigating these climate effects.

In Chapter 3, we obtained a first introduction to the causes of climate change by considering the Earth's energy budget. We need a much deeper insight into climate science to gain a useful understanding of the extreme climate conditions that are now emerging. Many of these can be understood by using a more elaborate model of the Earth's energy budget, particularly one that incorporates features of the Earth's climate system and the *Earth energy imbalance* referred to in Chapter 3 within a single framework.

Climate scientists depend on climate models to understand the current state of the Earth's climate and to predict its future, and a brief introduction to these indispensable tools is necessary to gain insight into the current climate research. These models are developed by several different modelling centres distributed all over the world, and the work of the World Climate Research Project in coordinating and rationalising their outputs towards the common cause is critical.

The climate impacts of most concern are those due to climate tipping points. It is therefore of vital interest to know what these are and the extent of the threat they pose to global society.

DOI: 10.1201/9781003531180-4

This chapter begins by introducing the work of *extreme event attribution (EEA)* groups who attribute observed climate impacts to climate change and proceeds to elaborate on the work of climate modellers who develop climate models that are now able to simulate the Earth's climate and provide deeper insights into human-induced climate phenomena.

It then proceeds with a brief discussion of the *forcing-feedback framework* that provides an integrated model to account for several climate impact phenomena within the single conceptual framework.

How the Earth's climate system is monitored and modelled is discussed next. The agencies and projects that monitor the Earth's climate system are enumerated. This is followed by an overview of the current generation of climate models and the modelling groups that maintain them. The chapter ends with an account of the nature of climate tipping points, the interactions between them, and the role of positive tipping points in strategies to achieve net-zero by 2050.

4.2 DOES CLIMATE CHANGE EXIST?

Chapter 2 provided a first introduction to the extreme weather events now facing the Earth. Before we can deliberate further into their causes, one must first ascertain that such weather anomalies are indeed due to human-induced global warming.

A simple method, conceived by Allen [1] in 2003, is to evaluate the contribution of global warming to the risk of an individual weather event by comparing the probability of an (extreme weather) event occurring to that of the same event occurring in a world with no human-made GHG emissions. The fraction of attributable risk (FAR) can then be assigned to the portion of the risk of the extreme weather event that is due to global warming. Note that this attribution can be either positive (global warming increases the risk), negative (reduces the risk) of the event occurring, or have no influence on the event. Such attributions have become known as *Extreme Event Attributions (EEAs)*.

In the same year, Europe experienced an extreme heat wave that is estimated to have killed some 70,000 people. Climate scientists Stott, Stone and Allen [2] used the above methodology to show that human-produced GHG emissions had at least doubled the risk of such an extreme heat wave occurring in Europe and demonstrated the value of EEA to the global community for the first time. Since then, several hundred EEA studies have been carried out on a similar number of extreme weather events to prove that such events are strongly correlated to human-induced climate change.

Carbon Brief has compiled an Attribution Map of more than 600 studies of around 750 extreme weather events [3] that reveals that

- 74% of the 750 events were made more likely or severe because of climate change.

- 9% were made less likely because of climate change.

- 10% had no human influence on them.

- 7% were inconclusive.

There are now dozens of groups working on EEA. One such group is the **World Weather Attribution** group [4]. This group was initiated by Dr. Oldenborgh and Dr. Otto in 2014 and now consists of a collaboration between the Grantham Institute, Imperial College of London, the Royal Netherlands Meteorological Institute and the Red Cross Red Crescent Climate Centre. An example of the work of WWA is a study of the extreme heat events of July 2023 in North America, Europe and China [5].

Their findings reveal that

- All these three extreme weather events that occurred concurrently in 2023 were made more likely because of climate change.

- The frequency of such events would increase dramatically if global warming were to reach 2°C.

Some of these results are summarised in Table 4.1.

TABLE 4.1   Some of the WWA Findings on the Global Extreme Heat Events of July 2023

Region	Likelihood of Heatwave with Climate Change	Likelihood in the Absence of Climate Change
North America	Expected once every 15 years	Impossible to occur
Europe	Expected once every 10 years	Impossible to occur
China	Expected once every 5 years.	Expected once every 250 years
Effect of warming temperature on likelihood	If global warming reached 2°C, heat waves similar to the July 2023 waves would occur once every 2–5 years.	

Source: [5].

4.3 THE FORCING – FEEDBACK FRAMEWORK

Three key concepts used by climate scientists to study the science of climate are *climate forcings, climate feedbacks* and *climate tipping points*. A fourth concept, related to climate feedback, is *climate sensitivity*.

NASA defines the first three of these as follows [6]:

- **Climate Forcings** are the initial driver of climate change, which include changes in solar irradiance (i.e. solar flux at the top of the atmosphere (TOA)), greenhouse gas (GHG) emissions, aerosols as well as dust, smoke or soot.

- **Climate Feedbacks** are processes that can either amplify or reduce the effects of climate forcings, and examples of these consist of clouds, precipitation, forest greening or browning, ice albedo and water vapour

- **Climate Tipping Points** are abrupt changes in the Earth's climate between relatively stable periods, as exemplified by changes in the (global) ocean circulation, changes in the permafrost found in the polar regions, and changes in ecosystems.

Climate sensitivity is a measure of how much warming will occur with an applied forcing. This information is invaluable in obtaining a first estimate of how much GHGs we can add to our atmosphere before the warming reaches the 1.5–2.0°C range specified by the Paris Agreement. An idealised case that is used by climate scientists is where the forcing is produced by a doubling of the Earth's atmospheric GHG concentration as compared to the pre-industrial levels.

Two definitions for the resulting warming are the **Equilibrium Climate Sensitivity (ECS)** and the **Transient Climate Response (TCR)**. ECS is the change (ΔT) in the value of the global surface air temperature (GSAT) when the radiative imbalance ΔN produced by the forcing has reduced to zero. TCR is the temperature change at the moment of doubling as the GHG concentration is increased gradually at the rate of 1% per annum [7].

The discussion above should indicate that climate change is a very complex process, and an understanding of the science behind climate change requires the use of a range of concepts that have been invoked for the specific task. In Chapter 3, we focussed only on the global warming due to a specific radiative forcing (and ignored the possibility of other forcings). In addition, the Earth Energy Budget (equation 3.3) did not include any reference to the forcing or the change (ΔN) in the radiative flux at the TOA. These aspects of the mechanism had to be introduced verbally to explain the mechanism of global warming.

It would clearly be of advantage if a better conceptual framework could be formulated that could integrate several of the new climate concepts within the same model, which can then be represented by a single equation. Such a generalised model is the *forcing-feedback framework* (also known as the *forcing and response energy budget framework*) reported by IPCC in its Sixth Assessment Report [8].

The model envisages a change in the net energy flux (ΔN) at the TOA produced by a forcing (such as an increase in the atmospheric GHG concentration) in terms of the linear energy budget equation

$$\Delta N = \Delta F + \alpha \Delta T \qquad (4.1)$$

where ΔF is the effective radiative forcing (ERF), α is the (climate) feedback parameter and ΔT is the change in the GSAT.

If the forcing is a doubling of the atmospheric CO_2 content as compared to its pre-industrial levels (often referred to as ERF for $2xCO_2$), the time evolution of expression 4.1 then enables the inclusion of ECS and the climate feedback parameter α within the same framework.

These relations are depicted in Figure 4.1.

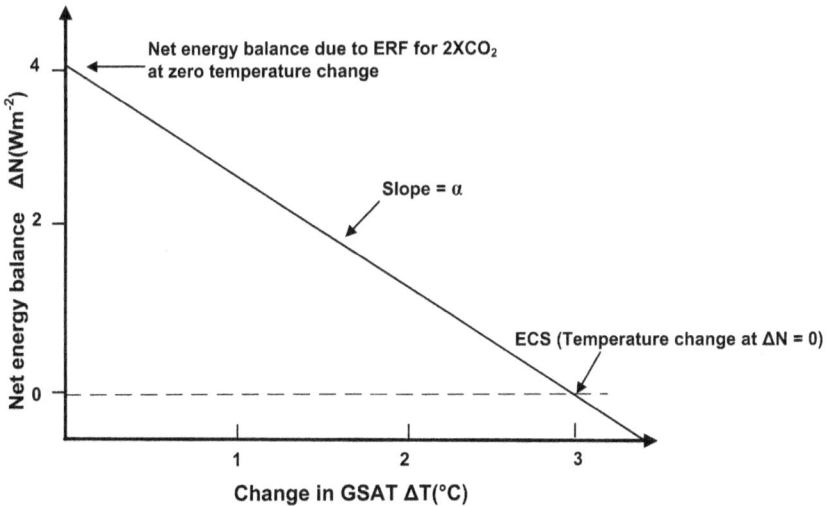

FIGURE 4.1 How the earth's net energy balance at the TOA evolves with time according to the forcing-feedback framework when the forcing is ERF for $2xCO_2$. The figure demonstrates how this framework is able to incorporate the concepts of ERF, climate feedback and ECS in an integrated manner. (Data source: [8]) (Figure credit: Asha Sinha.)

As seen from the figure, this framework is able to treat the notions of ERF, climate feedback and ECS in an integrated way by relating them all through the linear energy budget equation.

4.4 MEASURING AND MODELLING CLIMATE CHANGE

4.4.1 What Is the Earth's Climate?

Before embarking on modelling the impacts of climate change, it is instructive to look more closely at what is meant by climate.

The WMO defines climate as

> ... the average weather conditions for a particular location over a long period of time, ranging from months to thousands or millions of years [9].

It is best to begin by defining the Earth's climate system, which is made up of the [9]

- atmosphere,

- lithosphere (the land surface),

- hydrosphere (the oceans),

- cryosphere (ice and snow cover), and

- biosphere (living matter).

Climate is the interaction between all these components, through natural processes such as the exchange of energy, water and nutrients via the carbon and nitrogen cycles, as well as human-induced processes such as the emissions of GHGs and aerosols [9]. The state and dynamics of climate can be measured via climate variables such as temperature, humidity, precipitation (rain and snow), air pressure, solar irradiance and wind.

Climate is driven by energy from the sun, and its state may change due to both internal flows of mass and energy as well as external forcings. The latter consist of natural variations such as changes in the net solar flux, the earth's orbital, volcanic eruptions on the one hand, and human-induced forcings such as the release of GHGs and aerosols into the atmosphere and changes in land cover on the other.

4.4.2 Monitoring the Earth's Climate System

Experimental verification plays a crucial role in the advancement of good science, and climate science is no exception. Not only does it provide a test for the models and theories employed, but more importantly, it accords the invaluable opportunity for discovering new aspects of climate change not imagined before. A good example is the Earth energy imbalance discovered by the CERES project discussed in Chapter 3. The experimental discovery of this small "discrepancy" in the Earth's radiative energy balance led to the development of a whole new conceptual framework for climate science.

The Earth's climate system has been monitored by at least four international agencies and projects over the past several decades in programs to collect baseline and continuous data on developing trends in climate metrics. They include measurements from satellite, land, ship and balloon-based atmospheric observatories by agencies such as the NASA Earth Observing System (EOS), the National Oceanic and Atmospheric Administration (NOAA), the NASA CERES Project and the European Space Agency (ESA).

Below are brief introductions to these agencies and projects.

4.4.2.1 *The CERES Project*

The Clouds and the Earth's Radiant Energy System (CERES) project is a satellite-based project to gather data on the Earth's Radiation Balance (ERB) and clouds [10]. It measures reflected (short-wave solar) radiation and emitted (long-wave thermal) radiation at the TOA and combines this information with data from other sources to produce a comprehensive set of ERB data for science research.

The project uses seven instruments (radiometers) which were launched aboard five different satellites (TRMM, Aqua, Suomi National Polar-orbiting Partnership, and NOAA-20) [11].

The bandwidths investigated are

0.3–0.5 µm (short-wave solar)

0.3–200 µm (total)

5.0–35.0 µm (Long-wave thermal)

The project started collecting data in 1997.

4.4.2.2 NASA's Earth Observing System (EOS)

Initiated in the early 1990s, NASA's EOS program consists of a series of polar-orbiting satellites to monitor key features of the climate system using long-term monitoring [12]. Climate variables and elements investigated include radiation, clouds, water vapour and precipitation, the oceans, GHGs, land-surface hydrology and ecosystem processes, glaciers, sea ice and ice sheets, ozone and stratospheric chemistry and natural and anthropogenic aerosols.

Current EOS missions are Aqua, Terra, Aura and Landsat 7.

4.4.2.3 ESA's Climate Change Initiative and Missions

ESA's climate change initiative (CCI) was established in 2008 to develop global, decades-long satellite-derived climate data records for the IPCC and the climate science information "value chain" (including monitoring, modelling and climate services). Its goal from 2023 to 2029 is to provide the Earth Observation requirements to support the UNFCCC Paris Agreement and its implementation through the Nationally Determined Contributions (NDCs) and National Adaptation Plans (NAPs) [13].

The CCI generates datasets for key components of the climate, known as essential climate variables (ECVs) through more than 25 ECV projects. Examples are Aerosol, Biomass, Fire, Glaciers and Ice Sheets projects. Each of these projects elaborates on one or more of the ECVs, such as mole fractions of the GHGs and thickness and cover of ice sheets. Examples of current ESA Earth Observing Missions include Aeolus, Altius, artic weather satellite, biomass, chime, cimr, co2m and Copernicus Sentinel expansion missions [14].

4.4.2.4 NOAA's Global GHG Monitoring

NOAA, which is part of the US Department of Commerce, has an organisational mission to understand and predict changes in climate, weather, oceans and coasts, to share this information and to conserve and manage coastal and marine ecosystems and resources [15].

Towards these ends, it has established a Global Monitoring Laboratory in Boulder, Colorado. The Laboratory manages a Global Greenhouse Gas Reference Network that monitors atmospheric distributions of the three most abundant GHGs (CO_2, methane and nitrous oxide) as well as carbon monoxide at Atmospheric Baseline Laboratories and multiple tall towers in the US, for samples collected at 50 sites internationally by volunteers, in air samples collected regularly from small aircrafts in North America, and vertical profiles using balloons and aircore sampling systems.

The Laboratory also performs Annual Greenhouse Gas Index monitoring on samples collected from the GGGRN centres as well as ship routes at 5 degree latitude intervals [16].

4.4.3 What Are Climate Models?

Climate models are mathematical representations of the earth's climate that are used to predict the future of the earth's climate from knowledge of its past and present state. They are huge computer programs that (in their most basic forms) use the equations of physics to simulate the Earth's climate [17]. This essentially reproduces the processes, including energy transfer and mass flows, that result when the different components of the Earth's climate system interact with each.

The models accept climate variables that are of interest to climate scientists as inputs and produce outputs that constitute future climate scenarios that correspond to the inputs. Such scenarios provide crucial climate information to scientific organisations (such as the IPCC) that are interested in assessing how human-induced changes to our environment determines future climate.

Climate models may come either as Global Climate Models (also called General Circulation Models [GCMs]) that describe aspects of the global climate as a whole or as regional models that only consider specific regions of the Earth.

4.4.4 Types of Climate Models

Climate models come in a range of complexities and capabilities [18]. The simplest only do simple calculations, e.g. of the Earth energy balance, and treat the Earth as featureless, i.e. zero-dimensional. Examples are *Energy Balance Models (EBMs)* which simply calculate the Earth's energy balance and provide no other information pertaining to the Earth's climate.

Slightly more developed models are able to evaluate climate variables such as temperature and relative humidity along a single dimension, such as the height above the Earth's surface. The *Radiative Convective Models* are examples of such one-dimensional models.

Models at the next stage of sophistication are *GCMs*. Also called Global Climate Models, these produce three-dimensional simulations of the Earth's climate. They do this by breaking up the Earth's near-surface region into little cubes, and using the equations of physics to evaluate the energy and mass flow (e.g. wind or ocean currents) within each pixel.

Early GCMs were confined to modelling individual components of the Earth system (such as the atmosphere or the ocean) only. Later, these began to be combined into *Coupled GCMs* that provided more holistic perspectives of the Earth's climate. An example of a coupled GCM is an *Atmosphere-Ocean GCM (AOGCM)* that considers the atmosphere, the oceans and the interactions between them to produce a more holistic picture of the ocean-atmosphere climate. Carbon Brief [18] presents an infographic that shows how coupled GCMs have evolved from the individual component GCMs of the early 1960s to the comprehensive Coupled GCMs of the 2010s that now incorporate aspects of the Earth system such as the carbon and nitrogen cycles, sea ice and atmospheric chemistry.

Perhaps the most evolved type of GCMs is the *Earth System Models (ESMs)*. These now include the biogeochemical cycles, which make these systems capable of simulating the impact of human-induced changes to the environment and the Earth's climate. ESMs are capable of simulating processes such as ocean ecology and human-produced changes in vegetation and land use. These are all related to GHG emissions and therefore enhance the capability of ESMs to model the consequences of climate change.

Figure 4.2 presents a schematic outline of the evolution of climate models over time.

Climate modelling tools that have been developed to cater for general climate change applications include *Integrated Assessment Models (IAMs)* [19].

IAMs are tools that combine different strands of knowledge relating to human society and climate change that models how human socio-economic behaviours as a whole influence each other and the natural world. Inputs to these models may include the GDP, population and policies. The IAM produces outputs that are relevant to such socio-economic variables of a country. These outputs may consist of scenarios such as GHG emissions, energy and land use pathways that are appropriate to the inputs provided.

Such outputs are of relevance to the generation of climate scenarios used by the IPCC in identifying the most appropriate pathways for climate policy development. In particular, the IAM outputs can be used as inputs to ESMs to produce future climate scenarios that are needed by the IPCC. The reader is referred to Carbon Brief's post [19] for a comprehensive treatment that provides a better insight into these tools.

Climate models are developed and maintained at some 49 institutions scattered around the world. The names of the specific models they develop do not seem to follow any particular scheme. However, many of the model

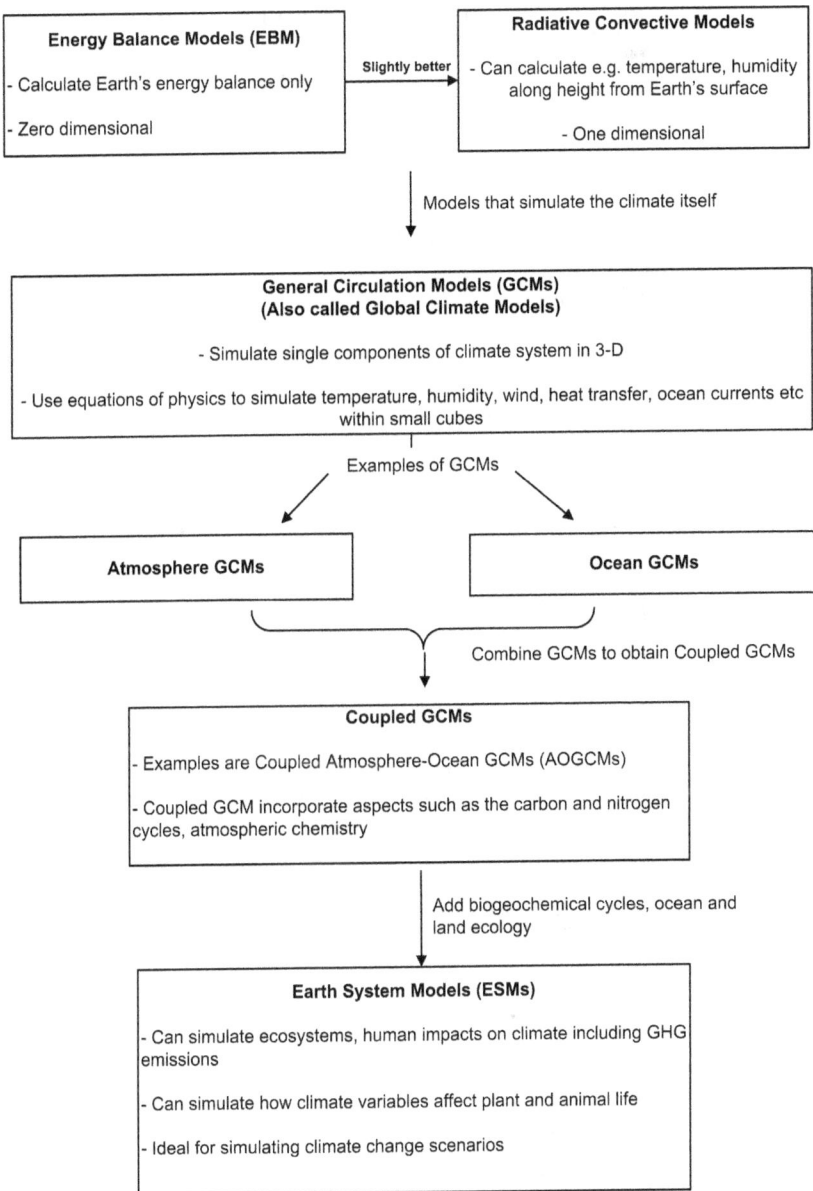

FIGURE 4.2 The evolution of climate models. (Figure credit: Asha Sinha.)

names start with name of the centre followed by the type of the model. For example, the model named *CanESM* refers to the ESM climate model developed by the Canadian Centre for Climate Modelling and Analysis [20].

Table 4.2 provides examples of some of the models and the centres that develop them.

TABLE 4.2 Examples of Climate Modelling Centres and the Climate Models They Develop

Region	Name of Climate Model	Name and Location of Centre
Europe	MP1	Max Planck Institute for Meteorology Hamburg, Germany
	EC-Earth	EC-EARTH Consortium Utrecht, Netherlands
	HadCM3, HadGEM2	Met Office, Hadley Centre Exeter, UK
	IPSL-CMS	Institut Pierre-Simon Laplace (IPSL) Paris, France
North America	CanESM, CanCM, CanAM	Canadian Center for Climate Modelling and Analysis (CCCMA) Victoria, Canada
	CESM1	Community Earth System Model Boulder, Colorado
	GFDL	NOAA Geophysical Fluid Dynamics Laboratory (NOAA GFDL) Princeton, USA
	GISS-E2	NASA Goddard Institute for Space Studies (NASA GISS) Washington DC, USA
	RSMAS, CCSM4	University of Miami (REMAS) Miami, USA
Asia	HadGEM2	National Institute of Meteorological Research/Kora Meteorological Administration (NIMR/KMA) Seoul and Jeju-do, South Korea
	FIO-ESM	The First Institute of Oceanography, SOA, China (FIO) Qingdao, China
South America	BESM-OA	National Institute for Space Research (INPE) Sao Paulo, Brazil
Australia	ACCESS	CSIRO/Bureau of Meteorology (BOM) Melbourne, Australia.

Source: [20].

4.4.5 Climate Models, the WCRP and the IPCC

An important use of climate models is in the production of climate projections for the *Intergovernmental Panel on Climate Change (IPCC)*, which needs this information to produce its Assessment Reports. This task is performed through a grand collaboration between the *World Climate Research Programme (WCRP)*, the climate modelling centres and the IPCC.

There are more than 49 climate modelling centres around the world, which currently use over 100 models to produce runs for the IPCC. The model outputs are used by the IPCC to prepare its assessment reports which climate policy-makers (the UNFCCC) use to develop their climate policies. As an example, an important modelling result required by the IPCC is how much the Earth will warm when a certain amount of GHG is added to the atmosphere. This is achieved by calculating the climate sensitivity when the forcing is caused by the human-induced change in atmospheric GHG concentration.

Generally, each modelling centre develops its own models. This means that the outputs from these models for the same input will generally be different. Thus, the combined ensemble of models will produce a range of possible climate futures for the world that the scientists at IPCC can consider. For the results to be useful, the manner in which the outputs from differing groups are produced must be standardised and carefully regulated. In particular, the inputs used during model runs for a specific IPCC report must be exactly the same for all models. Clearly, there is a need for coordination of the model runs. This task is performed by the WCRP that was established in 1980 [21]. This body is overseen by the *Joint Scientific Committee (JSC)* appointed jointly by the WMO, the *International Science Council (ISC)* and the *Intergovernmental Oceanographic Commission (IOC)*.

The coordination is carried out via a project known as the *Coupled Model Intercomparison Project (CMIP)* [22]. The first phase of this project (CMIP5) was used in the production of the *IPCC Fifth Assessment Report (IPCC AR5)*, while the second phase (CMIP6) coordinated the latest (sixth) report produced in 2021.

For more details about how the WCRP coordinates the work of the climate centres in the modelling of climate change for the IPCC, please see [23].

4.5 CLIMATE TIPPING POINTS

4.5.1 What Are Climate Tipping Points?

In Section 4.3, we saw that the new forcing-feedback framework was able to account for the first three of the four concepts (forcing, feedback, sensitivity and tipping points) used by climate scientists to understand climate change. So what conceptual tools do climate scientists use to investigate climate tipping points? To answer this, one must first know what tipping points are.

Climate tipping points are the temperature threshold at which parts of the earth's climate system, known as a *tipping elements*, change irreversibly and become self-perpetuating, i.e. remain in the new state even when the temperature falls below the threshold again [24, 25].

An example is the Antarctic sea-ice sheet, which grows during winter as the surface temperature falls, and contracts during summer as this temperature rises. If global warming increases beyond the tipping point in any year, the sheet may continue contracting even if the warming temperature falls below the threshold again. Here, the component of the climate system being considered is the cryosphere, and the ice sheet is the tipping element. The change in the tipping element may be abrupt and lead to environmental conditions that are unpredictable and harmful to life.

4.5.2 Detecting Climate Tipping Points

Is it possible to predict when a climate tipping point will occur? This is the same as predicting the warming temperature that will bring the tipping element to its threshold.

Dakos et al. have described the theory and use of the *early warning signals (EWS)* approach to detecting the onset of tipping points in climate, ecology and other disciplines [26]. This method is based on the detection of the slowing down of the system's recovery to its stable state as it evolves towards a tipping point. This *critical slowing down (CSD)* leaves signatures in the spatial and temporal dynamics that can be measured.

A practical (and intuitively more transparent) approach was used by Drijfhout et al. [27] for the detection of climate tipping points as early as 2015. These workers searched the data on model runs for the IPCC AR5 Report for abrupt shifts in tipping elements such as sea ice cover, permafrost, Boreal forest cover and Amazon forest cover after specific global temperature increases. They found that such shifts are significantly larger in runs using the RCP8.5 scenario as input. Table 4.3 provides a small sampling of their findings.

TABLE 4.3   Abrupt Changes in Tipping Elements in IPCC AR5 Model Runs

Category of Abrupt Change	Type of Abrupt Change	Region	Models and Scenarios
I (switch)	Sea ice bimodality	Southern Ocean	BCC-CSM1-1 (all)
II (Forced transition to switch)	Sea ice bimodality	Southern Ocean	GISS-E2-H (rcp8.5) GID-E2-R (rcp8.5)
III (Rapid change to new state)	Winter ice collapse	Arctic Ocean	MP1-ESM-LR (rcp8.5) CSIRO-MK3-6-0 (rcp8.5) HadGEM2-ES (rcp8.5)
	Abrupt sea ice decrease	Higher latitude ocean regions	CanESM2 (rcp8.5) CMCC-CESM (rcp8.5)
	Permafrost collapse	Arctic	HadGEM2-ES (rcp8.5)
IV (Gradual change to new state)	Boreal forest expansion	Arctic	HadGEM2-ES (rcp8.5)
	Forest dieback	Amazon	HadGEM2 (rcp8.5) IPSL-CMSA-LRC (rcp8.5)

Source: [27].

4.5.3 Types of Climate Tipping Points

4.5.3.1 *Earth System Tipping Points*

Climate tipping points can be conveniently divided into

i. physical tipping points, at which tipping elements of the *physical components* of the Earth's climate system undergo irreversible, self-sustaining change, and

ii. ecosystem tipping points, where tipping elements within the *biosphere component* of the climate system are at risk of undergoing similar changes.

Recently, the concept of *positive tipping points* has been mooted by prominent groups of scientists and economists interested in expediting solutions to the net-zero challenge (Ref [28, 29]. These relate to tipping points in the socio-economic domain and are defined in the next subsection. To distinguish between the tipping points relating to the earth's climate and the socio-economic tipping points, the former will be called by the specific name of Earth system tipping point wherever necessary.

A list of well-known examples of the (Earth system) tipping points is provided by ESA [30] and selected examples are presented in Table 4.4.

TABLE 4.4   Types of Climate (Earth System) Tipping Points

	Type of Tipping Point	Example	Impacts	Temperature Threshold
1.	Ice sheet collapse	West Antarctic ice sheet collapse (Antarctica)	Sea level rise	< 2°C
2.	Ocean current system collapse	Atlantic Meridional Overturning Circulation (AMOC) Collapse (North Atlantic)	Altered weather patterns, sea level rise,	> 4°C
3.	Boreal permafrost abrupt thaw	Boreal forest permafrost. (Canada).	Land surface → lake transformation. Methane release.	< 2°C
4.	Boreal Forest expansion/ dieback	Boral forest northwest expansion (Alaska)	Impact on biodiversity (Boreal forest/Tundra)	2–4°C
5.	Amazon rainforest dieback	The Amazon. S. America	Loss of vegetation, biodiversity. Carbon release.	2–4°C
6.	Coral reef die-off	Coral bleach (Indonesia, Australia).	Coral death. Loss of marine biodiversity	< 2°C

Source: [30].

In Table 4.4, the first three are examples of changes in tipping elements belonging to the cryosphere, ocean and the lithosphere and may be classified as physical tipping points. The latter three examples involve changes in ecosystems and are therefore ecosystem tipping points.

4.5.3.2 Positive Tipping Points

Positive tipping points may be defined (after [28]) as concerted human actions that enable the conditions for triggering rapid, large-scale desirable changes in society. In the context of climate action towards achieving net-zero emissions, examples of such tipping points include the recently observed acceleration in the sale of EVs and the exponential increase in renewable electricity.

Such tipping points may re-inforce each other to trigger cascades. For example, the acceleration in EV sales creates demand for batteries, leading to their increased production and rapid cost reductions. Thus, the EV tipping point results in the triggering of a tipping point in battery production.

The Global Tipping Points Report 2023 (GTP Report) [28] has ten key messages, amongst the most important of which are:

- Five Earth system tipping points are already on the verge of being crossed at the current level of global warming. These are the Greenland and West Antarctic ice sheets, warm-water coral reefs, North Atlantic Sub-polar Gyre Circulation and permafrost regions.

- These threats could materialise within the coming decades, at lower levels of global warming than previously thought.

- Triggering one Earth system tipping point could trigger another, causing a domino effect of accelerating and unmanageable damage.

- Stopping these threats is possible but requires urgent global action. Governance is needed across multiple scales to address the different drivers, and diverse, often irreversible, impacts of tipping points.

- Positive tipping points can accelerate a transformation towards sustainability.

- Positive tipping points require coordinated action to create the enabling conditions needed for their triggering.

- Positive tipping points can create a powerful counter-effect to the risk of Earth system tipping points cascading out of control.

The topic of positive tipping points will be discussed at length in Chapter 9.

4.5.4 Interactions between Climate Tipping Points

The report quoted above mentions the possibility of the occurrence of cascades of Earth system tipping points similar to the positive tipping point cascades. The first requirement for this to occur is an interaction between the various climate tipping points, and it is interesting to see how such interactions are possible.

A review has been carried out of such interactions by Wunderling et al. [31]. They find that such interactions are indeed possible, and that

- The causal link between tipping elements can be stabilising, destabilising or unclear

- Many of the interactions between tipping elements are destabilising

- Tipping cascades cannot be ruled out on long timescales (centennial to millennial) at warming levels between 1.5 and 2.0°C

- At levels of warming that are higher than 2°C, tipping cascades may involve fast tipping elements such as the AMOC and the Amazon rain forest.

4.5.5 Impacts of Climate Tipping Points

To better understand the impacts of climate tipping points, one needs a detailed model of the process that occurs during the tipping point period. Such a model is provided by ESA [30].

This pictures the climate system in terms of a landscape having two stable states. (Alternatively, the system may be viewed as a physical object such as a ball rolling down a slope which has two consecutive wells downhill).

Initially, the system is in state 1 (or the ball is in the first well). Global warming changes the landscape (the potential well seen by the ball) so that the barrier separating the two states (the little hillock separating the wells) is reduced. At the tipping point, the system crosses over to state 2 (the ball rolls over to the second well since the hillock has disappeared due to global warming). The climate system has now crossed the temperature threshold of the tipping point.

The question that needs to be answered is how the second state compares with the first. This is the same as asking what the consequences of crossing the climate threshold are. Table 4.4 lists some of the impacts of the Earth System

TABLE 4.5   An Overview of the Earth System Tipping Points and Their Impacts as Reported in the GTP Report

Earth System Tipping Point	Result of Crossing the Tipping Point
General	Crossing the tipping point would have severe impacts on people and biodiversity, including water, food and energy insecurity, and disruptions to social cohesion, escalating conflicts and economic insecurity.
Amazon dieback	• Catastrophic effect on biodiversity • Enhanced global warming • Extreme heat stress to humans
Antarctic ice sheet instability	• Sea level rise of up to 2 metres by 2100 • 480 million people at risk of annual coastal flooding events
Permafrost thaw	• Enhanced global warming • Loss of land area for human habitation • Damage to property and infrastructure
AMOC collapse	• Reduction in (food and energy) crop production over large land areas of the world • Severe implications for food security

Source: The GTP Report [28].

tipping points and the temperature threshold at which they occur. A more comprehensive and illustrative review of these impacts has been reported in the GTP Report [28]. Table 4.5 provides a very brief overview of these impacts.

As is obvious from Table 4.5, the impacts of tipping points are more far-reaching than previously imagined and transcend socio-economic boundaries to pose hazards to society and the biosphere as a whole. They will be subjects of central interest in the last chapter.

4.6 SUMMARY

1. A simple way of demonstrating that extreme weather events such as wildfires and heatwaves are caused by climate change is to compare the probability of such events occurring to that if they happened in a world without human-made GHGs. Such EEAs have been made by hundreds of scientific groups. In a study of 750 extreme weather events, it was found that 74% were made more likely or severe because of climate change.

2. The science of climate change uses several key concepts, which include climate forcings, feedbacks, sensitivity and tipping points. A conceptual framework that relates the first three of these together is the forcing-feedback framework, which plays a central role in the

understanding of how human-produced perturbations to the Earth's energy balance produce climate change.

3. Climate is the average weather conditions (rainfall, temperature, wind, sunshine, ocean current, salinity, etc.) for a particular location over a long period of time. It is driven by energy from the sun and determined by the interactions between the components of the Earth's climate system (atmosphere, land surface, ocean, ice cover, biosphere).

4. Over the past several decades, the Earth's climate system has been monitored from the Earth and satellites from above by at least four global agencies and/or projects. These are the CERES project, NASA's EOS, ESA's CCI and missions and NOAA's Global GHG Monitoring project.

5. Climate models are (usually) large computer programs that simulate the Earth's climate using the equations of physics and the other basic sciences. These programs accept climate variables (such as temperature, humidity, wind velocity) as inputs and produce ECVs as outputs of interest to climate scientists involved in climate action.

   The simplest climate models are the EBMs. GCMs are much more complex and provide 3D simulations of the Earth's climate by calculating the climate variables within small cubical elements of specific components of the climate system. These GCMs may be coupled together into Coupled GCMs to produce outputs for two or more climate components and the interactions between them. The most sophisticated models are the ESMs, which simulate biogeochemical processes and are able to model future climate scenarios.

6. Climate models are developed by more than 49 modelling centres around the world. These model outputs are used by the IPCC in the preparation of their assessment reports. There is a need to coordinate the work of these various modelling groups so that the outputs are standardised. This coordination is carried out by the World Climate Research Program (WCRP), under the Coupled Model Intercomparison Project (CMIP). The first phase (CMIP5) was used to coordinate the production of the IPCC AR5 report, and the second (CMIP6) was used for the IPCC AR6 report.

7. Climate tipping points are the temperature threshold at which tipping elements within the Earth's climate system change irreversibly and the change becomes self-perpetuating. Climate tipping points

(also called Earth system tipping points) may be conveniently divided into physical and ecosystem tipping points.

8. A new type of tipping point that is used in relation to the energy transition is a positive tipping point. This relates to concerted human actions that trigger rapid and desirable changes in technology towards the achievement of the net-zero goal. An example is the exponential increase in the sale of EVs. The Global Tipping Points Report 2023 shows how positive tipping points can be used to accelerate progress towards achieving net-zero emissions by 2050.

9. Climate tipping points can interact with one another to trigger cascades of tipping points. These become more probable at levels of warming above 2°C.

10. Crossing climate tipping points will have severe impacts on people and biodiversity, including water, food and energy insecurity. For instance, crossing the Amazon dieback tipping point will produce catastrophic effects on biodiversity and extreme heat stress to humans. An AMOC collapse will reduce food and energy crop production over large land areas of the world and have severe implications on food security.

REFERENCES

1. Allen, M. Liability for climate change. Nature 421 (2003) 891–892.
2. Stott, P., Stone, D., & Allen, M. Human contribution to the European heatwave of 2003. Nature 432 (2004) 610–614.
3. CarbonBrief. Just published…Carbon Brief's "attribution map" update and in-depth Q&A. Available from https://mail.google.com/mail/u/0/?pli=1#inbox/WhctKLbNDPqRnhgXcrkfHsgLmp VNlVXBPFjfZDZRZlrxRZV PlQRdMdHGtGPLJrchzsRvtxq. Accessed 16 Dec 2024.
4. World Weather Attribution. FAQs. Who is world weather attribution?. Available from https://www.worldweatherattribution.org/faqs/. Accessed 16 Dec 2024.
5. World Weather Attribution. Extreme heat in North America, Europe and China in July 2023 made much more likely by climate change. 25 July 2023. Available from https://www.worldweatherattribution.org/extreme-heat-in-north-america-europe-and-china-in-july-2023-made-much-more-likely-by-climate-change/. Accessed 16 Dec 2024.
6. NASA Global Climate Change. The study of Earth as an integrated system. Available from https://climate.nasa.gov/nasa_science/science/. Accessed 22 Nov 2024.
7. UK Met Office. What is climate sensitivity? Available from https://www.metoffice.gov.uk/research/climate/understanding-climate/climate-sensitivity-explained. Accessed 23 Nov 2024.

8. Forster, P. et al. The Earth's energy budget, climate feedbacks and climate sensitivity. In Climate Change 2021: The Physical Science Basis. Contribution of Working Group 1 to the Sixth Assessment Report of the IPCC. CUP. Ch 7. Available from https://www.ipcc.ch/report/ar6/wg1/chapter/chapter-7/. Accessed 22 Nov 2024.

9. WMO. Climate. Available from https://wmo.int/topics/climate. Accessed 3 Nov 2024.

10. NASA CERES. What is CERES?. Available from https://ceres.larc.nasa.gov/#:~:text=The%20Clouds%20and%20the%20Earth%E2%80%99s%20Radiant%20Energy%20System,products%20for%20climate%2C%20weather%20and%20applied%20science%20research. Accessed 23 Nov 2024.

11. NASA CERES. CERES instruments. Available from https://ceres.larc.nasa.gov/instruments/. Accessed 23 Nov 2024.

12. NASA's Earth Observing System. Missions: Earth observing system (EOS). Previously Available from https://eospso.gsfc.nasa.gov/mission-category/3. Accessed 23 Nov 2024.

13. ESA Climate Office. Overview of the climate change initiative. Available from https://climate.esa.int/en/about-us-new/climate-change-initiative/Climate-Change-Initiative/. Accessed 25 Nov 2024.

14. The European Space Agency. Earth observing missions. Mission navigator. Available from https://www.esa.int/Applications/Observing_the_Earth/Earth_observing_missions. Accessed 25 Nov 2024.

15. NOAA. US Department of Commerce. About our agency. Available from https://www.noaa.gov/about-our-agency. Accessed 25 Nov 2024.

16. NOAA Global Monitoring Laboratory. Annual greenhouse gas index (AGGI). Available from https://gml.noaa.gov/aggi/aggi.html. Accessed 25 Nov 2024.

17. USDA Climate Hubs. US Department of Energy. Basics of global climate models. Available from https://www.climatehubs.usda.gov/hubs/northwest/topic/basics-global-climate-models. Accessed 15 Dec 2024.

18. CarbonBrief. Q&A: How do climate models work?. Available from https://www.carbonbrief.org/qa-how-do-climate-models-work/#types. Accessed 26 Nov 2024.

19. CarbonBrief. How integrated assessment models are used to study climate change. Available from https://www.carbonbrief.org/qa-how-integrated-assessment-models-are-used-to-study-climate-change/. Accessed 26 Nov 2024.

20. CarbonBrief. Q&A: How do climate models work? Available from https://www.carbonbrief.org/qa-how-do-climate-models-work/#who. Accessed 15 Dec 2024.

21. World Climate Research Programme. About us. Available from https://www.wcrp-climate.org/about-wcrp/wcrp-overview. Accessed 29 Nov 2024.

22. CMIP. CMIP overview. Available from https://wcrp-cmip.org/cmip-overview/. Accessed 29 Nov 2024.

23. USDA Climate Hubs. US Department of Agriculture. FAQs about climate models. Available from https://www.climatehubs.usda.gov/hubs/northwest/topic/faqs-about-climate-models. Accessed 29 Nov 2024.

24. CarbonBrief. Global warming above 1.5°C could trigger 'multiple tipping points'. Available at https://www.carbonbrief.org/global-warming-above-1-5c-could-trigger-multiple-tipping-points/. Accessed 15 Dec 24.

25. Armstrong McKay, D. et al. Exceeding 1.5 C global warming could trigger multiple climate tipping points. Science 377 (6611) (9 Sept 2022). Available from https://www.science.org/doi/epdf/10.1126/science.abn7950. Accessed 15 Dec 24.

26. Dakos, V. et al. Tipping point detection and early warnings in climate, ecological, and human systems. European Geosciences Union. ESD 15 (2024) 1117–1135. Available from https://esd.copernicus.org/articles/15/1117/2024/. Accessed 1 Dec 2024.

27. PNAS. Drijfhout, S. et al. Catalogue of abrupt shifts in intergovernmental panel on climate change climate models. 112 (43) (2015) E5777–E5786. Available from https://www.pnas.org/doi/full/10.1073/pnas.1511451112. Accessed 1 Dec 2024.

28. Lenton, T. M. et al. (eds.) 2023. The Global Tipping Points Report 2023. University of Exeter, Exeter, UK. Available from https://global-tipping-points.org/resources-gtp/. Accessed 11 Dec 2024.

29. World Economic Forum. Climate action. Positive tipping points: a credible way to meet climate and nature goals. 8 July 2024. Available from https://www.weforum.org/stories/2024/07/positive-tipping-points-climate-nature-goals-wef/. Accessed 11 Dec 2024.

30. ESA. Understanding climate tipping points. Available from https://www.esa.int/Applications/Observing_the_Earth/Space_for_our_climate/Understanding_climate_tipping_points. Accessed 12 Dec 2024.

31. Wunderling, N et al. Climate tipping point interactions and cascades: A review. Earth System Dynamics 15 (2024) 41–74. Available from https://esd.copernicus.org/articles/15/41/2024/esd-15-41-2024.pdf. Accessed 11 Dec 2024.

CHAPTER 5

Climate Action and the Paris Agreement

5.1 INTRODUCTION

Chapters 3 and 4 described the science behind the cause and impacts of climate change. This chapter describes the global policies and implementation plans for addressing the challenge of climate change. It provides an overview of the method mandated by the global community, in the form of the Paris Agreement, to deal with the problem, describes how *Nationally Determined Contributions (NDCs)* are used to implement the Articles of the Agreement and assesses the overall success of global action so far in addressing the climate issue.

The chapter begins with a brief history of the *United Nations Framework Convention on Climate Change (UNFCCC)*, the governing body appointed by the UN to pursue action to address climate change, and introduces the key policies contained in the Articles of the Paris Agreement reached in Paris in 2015. These policies are implemented through the device of the NDCs, which are introduced next.

The original document of the Paris Agreement did not contain the details of the procedures and processes required for the operationalisation of the Agreement. These have been developed over the years and are known collectively as the *Paris Rulebook*. This is described in Section 5.5 with relevant examples. How effective the meetings of the UNFCCC (called COPs) have been is discussed next. This consists of a list of outcomes of the COPs, followed by and an assessment of their efficacies by two independent agencies.

DOI: 10.1201/9781003531180-5

The chapter ends with the outcomes of COP29 held at Baku in 2024 and considers the way ahead for future COPs.

5.2 HISTORY OF CLIMATE POLICY

Global climate policy is overseen by the UNFCCC. This is a multilateral treaty amongst 198 countries of the world, with the objective of

> stabilization of greenhouse gas concentrations in the atmosphere at a level that would prevent dangerous anthropogenic interference with the climate system …. within a timeframe sufficient to allow ecosystems to adapt naturally to climate change, to ensure that food production is not threatened, and to enable economic development to proceed in a sustainable manner [1].

It was formed at the *UN Conference on Environment and Development (UNCED* – also known as the *Earth Summit)* in Rio de Janeiro held from 3 to 14 June 1992, and came into force on 21 March 1994. The formation of the UNFCCC was motivated in part by the release of the First Assessment Report (FAR) of the IPCC in 1990, which drew attention to the rising concentrations of GHGs in the atmosphere [2].

The report explicitly revealed, for the first time, that

- There is a natural greenhouse effect that already keeps the Earth warmer than it would otherwise be.

- Emissions resulting from human activities are substantially increasing the atmospheric concentrations of the greenhouse gases (GHGs): carbon dioxide, methane, chlorofluorocarbons (CFCs) and nitrous oxide.

It also noted that these increases would enhance the greenhouse effect, resulting on average in an additional warming of the Earth's surface.

The IPCC used four scenarios for future emissions to predict future climate. It is remarkable that the first assessment was able to predict global temperature increases to lie in the range of 1.5–4.5°C by 2025–2050, and sea level rise of about 0.3–0.5 m by 2050 and 1m by 2100.

As noted above, the UNFCCC is a *treaty*, which is essentially a binding agreement amongst states on a particular issue. The UNFCCC binds states (countries) that are members of the UN and are party to the treaty on the issue of climate change. Members of the UNFCCC are called *Parties* (in 2024 there were 198 Parties). They meet at annual conferences called *Conference of the Parties (COPs)*, organised by an organising committee known as the *Secretariat*. The first COP was held in Berlin in 1995.

As the development of the UNFCCC was guided by the recommendations of the *FAR* of the IPCC, a clearer insight into the history of its formation is thus obtained by noting the details of the IPCC's report. This report began by noting that the world's energy sector accounted for 46% of the human-made GHG emissions, and that 70%–90% of total emissions arose from fossil fuels. Thus, a major part of the emissions originated from the industrialised countries, which therefore had to bear most of the responsibility. The report also noted that developing countries would need to industrialise and produce additional emissions in the process. To keep such emissions to a minimum, there was a need for technology transfer from the developed to the developing countries.

All this amounted to a call to the industrialised world to "adapt their domestic measures to limit climate change by adapting their own economies". They were also requested to assist developing nations in limiting their emissions through appropriate technology transfer.

These requirements played a pivotal role in the evolution of the UNFCCC towards arriving at the Paris Agreement.

Since its inception, the UNFCCC has had 29 meetings or COPs, the last meeting (COP29) being held in Baku, Azerbaijan, in 2024. Box 5.1 presents a chronology of the outcomes of some of the key COPs, assembled from various sources [3, 4].

BOX 5.1   A BRIEF HISTORY OF KEY UNFCCC COPs

COP1 – BERLIN 1995

This was guided by recommendations from the Subsidiary Body for Scientific and Technical Advice, the Intergovernmental Negotiations Committee and the IPCC. Commitments for emissions reductions were being sought from the industrialised (i.e. developed) countries (also called the Annex 1 countries) only, which were considered to be the largest emitters. Though certain Annex 1 countries expressed dissatisfaction with this "lack of universal commitments" [4], the Parties agreed to submit their national communications for anthropogenic emissions reductions.

COP2 – GENEVA 1996

There was discontent from the developed countries for being the only Parties required to commit to emissions reductions. In the end, Parties agreed to set binding emissions targets by industrialised countries (the

Northern States). The results of the IPCC Second Assessment Report were endorsed.

COP3 KYOTO, JAPAN 1997

At COP2, the Parties had been unsuccessful in establishing legally binding and quantitative targets for emissions reductions. COP3 was successful in adopting the *Kyoto Protocol*, which selectively committed the developed countries to emissions reduction. The US however did not participate in the agreement till a later COP. The developed nation Parties signed an agreement to reduce emissions to 5.2% below 1990 levels over the 2008–2012 period. The Kyoto Protocol also established carbon markets.

COP4 BUENOS AIRES 1998

This adopted the Buenos Aires Action Plan, which established rules for market-based mechanisms (e.g. emissions trading, joint implementation (JI) and clean development mechanism (CDM). It also developed compliance rules.

COP13 BALI 2007

Parties agreed to the *Bali Roadmap*, which included the *Bali Action Plan*, detailing "a new comprehensive process to enable the full, effective and sustained implementation of the Convention through long-term cooperative action". According to this plan

- Developed nations were to commit to "quantified emission limitation and reduction".
- Developing nations were to take mitigation actions "supported and enabled by technology, financing, and capacity-building".

COP15 COPENHAGEN 2009

Talks on a deal to replace the Kyoto Protocol (due to expire in 2012) with a global climate deal took place in an atmosphere of disappointment and distrust, sparked by a leaked document dubbed the "Danish text", which, amongst other things, gave effective control of climate finance to the World Bank. A nominal deal, with no mention of legally binding commitments, was eventually struck. However, this did include a mention of the 2°C warming limit.

COP16 CANCUN 2010

The highlight of this COP was the creation of the Green Climate Fund to finance developing countries' climate action.

COP18 DOHA 2012

It was decided to extend the Kyoto Protocol to 2020. This decision, however, did not receive unanimous support.

COP21 PARIS 2015

The Paris Agreement, with the goal of limiting warming to 2°C and making efforts to keep it within 1.5°C, was adopted. Its ratification however had to await another year.

COP22 MARRAKESH 2016

The Paris Agreement was ratified and came into force a few days before the start of this COP.

The Marrakesh Action Proclamation was made (in light of the imminent instatement of Donald Trump as the American President). It had a strong political message in support of the Paris Agreement.

COP23 BONN 2017

Progress was made on the Paris Rulebook, the rules by which the Paris Agreement was to be operationalised, with a view to completion by 2018.

As is apparent from Box 5.1, a key challenge in the development of the UNFCCC's goal was determining the relative roles of the developed nations (or Annex 1 nations) and the developing nation Parties. The first legally binding commitment to emissions reductions was the Kyoto Protocol, which required only the Annex 1 countries to commit to emissions reductions, starting with a target of 5.2% below 1990 levels. There was subsequently a concerted drive by these Parties to make the developing Parties to also commit somehow to emissions reductions. This eventually resulted in the adoption of the *Bali Roadmap* at COP13 in 2007, which committed the latter Parties to take mitigation actions with (technological and financial) support [5, 6].

How the COPs have performed is discussed briefly in Section 5.6.

5.3 THE PARIS AGREEMENT – KEY ARTICLES

The Paris Agreement consists of 29 Articles arranged in paragraphs [7]. The main thrust of the Agreement is contained in articles 2 to 14, which describe the objectives and their implementation through NDCs by each Party towards global emissions reductions. The remainder of the Articles are essentially about operational issues pertaining to the initiation of the Agreement. Table 5.1 presents the key Articles.

TABLE 5.1   Key Articles of the Paris Agreement

Article no.	Article Theme	Objective/Methodology
2.	Establish temperature limits to global warming	a. Hold global warming to well below 2°C above pre-industrial levels and make efforts to limit increase to 1.5 °C b. Make finance flows available for a pathway towards low emissions.
4.	Limiting and reducing national annual emissions of GHGs through NDCs	1. Parties should aim at global peaking of emissions of greenhouse gases (GHGs) as soon as possible (after 2015) and a rapid decline thereafter. 2. Each Party should prepare successive Nationally Determined Contributions (NDCs) to its annual GHG emissions reductions and communicate it to the Secretariat for recording in the public registry. 3. Support will be provided to developing nations to implement this Article. 4. All Parties to formulate and communicate long term Low Emissions Development Strategies (LEDs), taking into account their common but differentiated responsibilities and respective capabilities.
5.	Using forests as sinks	To use forests as carbon sinks, prevent de-forestation and practice sustainable management of forests.
6.	Voluntary contributions towards NDCs involving carbon trading.	Use of voluntary schemes, including internationally transferred mitigation outcomes (ITMOs) such as carbon trading/credits, towards enhancing a Party's NDC.
7.	Adaptation and vulnerability to climate impacts	Parties to enhance action on adaptation according to the Cancun Adaptation Framework.
8.	Loss and damage due to climate change impacts	To address loss and damage due to climate change impacts in accordance with the Warsaw International Mechanism for Loss and Damage.
9.	Financing developing countries	Developed country Parties will provide financial resources to developing countries for both mitigation and adaptation.
10.	Climate change technology development and transfer	Parties will strengthen cooperative action on technology development and transfer. Support to the developing countries will be provided for the implementation of this Article.
14.	Global stocktake	The COP to undertake periodic global stocktake of the implementation of the Paris Agreement. The first stocktake to take place in 2023 and then one every five years.
15.	Implementation committee	The COP to establish an expert committee to facilitate the implementation and compliance of the Paris Agreement.

Source: [7].

These Articles provide a comprehensive policy framework for global action to address climate change.

5.4 THE NATIONALLY DETERMINED CONTRIBUTIONS (NDCs)

The NDCs provide one of the key methodologies for implementing the Articles of the Paris Agreement. What they are, and how they work, is described below.

5.4.1 What Are the NDCs?

The NDCs are the contributions Parties are required to make towards the reduction of (annual) global GHG emissions to hold global warming within the temperature limits described in Article 2 of the Agreement. They are introduced in Article 3 of the Agreement and described more fully in Articles 4 and 6.

NDCs are the main policy mechanism for the implementation of the Agreement. An NDC is the action plan of a Party to achieve its share of emissions reductions and to adapt to climate change impacts [8]. NDCs have short-to-medium term targets and are required to be updated every five years, with each update more ambitious than the previous NDC [9].

NDC targets may be either *conditional* or *unconditional*. NDC targets that a Party can achieve on its own (i.e. through its own resources) are called *unconditional targets*, while *conditional targets* are those that can be achieved only through external support. All industrialised countries, as well as other high-income countries such as South Korea, Mexico and China submit unconditional NDCs only. All developing/low-income countries submit unconditional as well as conditional NDCs.

5.4.2 How the NDCs Work

- Each Party prepares its NDC and submits it to the Secretariat for recording in the public registry available at [10].

- This NDC is updated every 5 years, with the next update being due in 2025 (called NDC3.0) [8].

- To estimate its annual emissions reductions, a Party must first determine its total annual emissions from all its emissions sectors. These are [11]

  - Energy

  - Industrial Processes and Product Use (IPPU)

- Agriculture, Forestry and Other Land Use (AFOLU)

- Waste

- Other

- The emissions estimates are usually carried out according to the IPCC 2006 Guidelines and their 2019 revision [11].

The emissions mentioned above are measured in metric tons of "CO_2 equivalent", which indicates the mass of GHGs in a scheme where the mass of each GHG is weighted by its relative GHG efficacy (called Global Warming Potential) as compared to carbon dioxide. As an example, the US emitted approximately 6.34 Gigatons of CO_2 equivalent in the year 2022, of which 79.7% was due to CO_2, 11.1% to CH_4 and 6.1% to N_2O [12]. The percentages for CH_4 and N_2O were evaluated in this estimation after their respective masses were multiplied by their global warming potentials (28 and 265, respectively).

How all this is related to real-world activities can be understood by noting that emissions are produced by economic activities, such as fossil fuel-powered electricity generation and transportation. Thus, reductions in GHG emissions will be brought about by reducing the use of such sources of emissions in the economy (typically by replacing them with less-emitting alternatives). The implementation strategies for such reductions are called *net-zero strategies* and will be the subject of the next chapter.

5.5 OPERATIONALISING THE PARIS AGREEMENT – THE PARIS RULEBOOK

The Paris Agreement provides global policies that aim to deliver the climate goals of the UNFCCC. To implement these in practice requires tools, processes and guidelines that together form the action plan for the necessary climate action. These are compiled into "Rulebooks". Such rulebooks were not provided with the Agreement in 2015 but were developed over several COPs later, in particular COP24 held in Katowice, Poland in 2018.

An example is the rulebook relating to the NDCs (see Article 4 of the Agreement) [13].

This requires a Party to prepare successive NDCs and communicate them to the public registry managed by the Secretariat. However, it does not provide any detailed guidelines for what is to be submitted and the how this is to be done. This information is needed for the *clarity, transparency and understanding (CTU)* of the submitted, and the associated rulebook provides this information.

The absence of such information caused confusion in the format and content of the first NDCs submitted. For instance while some Parties submitted 5-year plans, others produced plans for longer periods and target dates that were further in the future. This led to information gaps in the submissions.

Another example relates to the accounting of emissions produced by a Party and the establishment of baselines (the year from which the emissions reductions were counted) [13].

At COP24:

- Negotiations on the NDCs were successful in producing a detailed list of required information to be included in the rulebook. These included

  - How to account for emissions and removals that are in accordance with methods and metrics approved by the IPCC.

  - How to establish baselines for emissions that are consistent between the communication and implementation of the NDCs.

  - Categories of emissions to be included.

COP24 was not successful in reaching agreements on rules for all the key Articles of the Agreement. One issue that could not be agreed on was the important issue of cooperative implementation (of the Paris Agreement) [13]. The three separate elements for which agreement could not be reached related to Article 6 (voluntary contributions to the NDCs). Specifically, these related to

- The use of internationally transferred mitigation outcomes (ITMOs),

- Emissions reductions credits (i.e. *carbon credits*) that could be used towards a Party's NDC, and

- Non-market approaches to implementing an NDC.

Rules for these were finally agreed to at COP29 in 2024 (see Section 5.7 for details). This finally saw the completion of the Paris Rulebook.

5.6 ACHIEVEMENTS OF THE COPs

The achievements of the COPs are essentially contained in the list of their outcomes. But to rigorously assess how successful the UNFCCC has been in its action to address climate change, it is necessary to compare the outcomes with relevant indicators of success. This section begins by elaborating on the indicators before presenting the outcomes. These are followed by their independent assessments carried out by the United Nations Environmental Programme (UNEP) and the World Meteorological Organisation (WMO).

5.6.1 Indicators of Success

How can we know how successful the global climate action has been?
  Success is best assessed from two independent perspectives:

- Action taken by the UNFCCC itself – these are the outcomes of the COPs

- Use of independent adjudicators who

  - Evaluate the state of emissions reduction, an example being the UNEP and its annual Emissions Gap Reports (EGRs)

  - Provide metrics on the state of global warming and its impacts on the Earth's climate, the leading authority being the WMO which produces its annual *State of the Climate Report* as well as other reports.

  Section 5.6.3 elaborates on the work done by the UNFCCC, UNEP and WMO in this regard.

5.6.2 Outcomes of the COPs

Climate change impacts are caused by global warming, and therefore the most critical results are achieved through limiting warming through mitigation actions. As nearly three-quarters of emissions result from the use of fossil fuels, this translates into reduction in the use of fossil fuels, which largely comprise coal, oil and gas. However, it was not until COP26 in 2021 that any direct mention was made of plans to reduce such emissions. The agenda of the COP26 meeting included aims to [14, 15]:

- accelerate the phase-out of coal

- curtail deforestation

- speed up the switch to electric vehicles

- encourage investment in renewables.

The final outcomes of the meeting, contained in the Glasgow Climate Pact, included a call to "phase down unabated coal power and inefficient subsidies for fossil fuels".

The *COP27* meeting, held in Egypt in 2022, merely re-iterated the COP26 sentiment by agreeing to

> accelerate efforts towards the phasedown of unabated coal power and phase-out inefficient fossil fuel subsidies.

COP28, held in Dubai, UAE between November and December 2023, made little progress beyond this, managing only to acknowledge the lack of progress made at that COP in phasing out fossil fuels, and included

> a call on governments to **speed up the transition away from fossil fuels** to renewables such as wind and solar power in their next round of climate commitments.

In summary, the outcomes of COP28 added little to the call to phase down unabated coal power made in COP26. This led to the results that, by the year 2023, the Parties making up the UNFCCC had not succeeded in making adequate commitments (via their NDCs or otherwise) to reducing emissions sufficiently to ensure the achievement of net-zero by 2050. This is clearly revealed in the assessments carried out by independent assessors, as shown below.

5.6.3 Independent Assessments of Achievement

The above outcomes relate to meetings of the UNFCCC. In effect, they are an assessment by the UNFCCC of its own performance towards achieving net-zero. The *United Nations Environment Programme (UNEP)* and the *WMO* are two agencies that provide independent assessments of the achievements of climate action undertaken by the UNFCCC.

5.6.3.1 *Independent Assessment by UNEP*

> UNEP reports annually on the status of the Emissions Gap through Emissions Gap Reports (EGRs). An emissions gap is essentially the difference between what the emissions should be

in 2030 to achieve the Paris Agreement goal of net-zero by 2050, and the projected emissions in 2030 under the current emissions in any year. The latter emissions can be calculated from the NDCs of the Parties as reported by them in the public registry.

Table 5.2 indicates the status of the emissions gap over three consecutive years in the 2022–2024 Period and Their Implications on the Respective NDCs.

Table 5.2 clearly shows that the world has been lagging far behind in its efforts to reduce global emissions sufficiently to reach net-zero by 2050. This condition is needed to ensure that global warming is contained within at most 2°C of pre-industrial levels by the end of 2100.

TABLE 5.2   Key Messages of Recent UNEP Emissions Gap Reports on the Status of the Gap between the NDCs and the Reductions Required by the Paris Agreement

UNEP Emissions Gap Report (EGR)	Assessment
EGR 2022 – Closing the window	• Policies currently (2022) in place point to a 2.8°C temperature rise by the end of the century. • Implementation of the current (NDC) pledges will only reduce this to a 2.4–2.6°C temperature rise by the end of the century, for conditional and unconditional pledges, respectively. • Only an urgent system-wide transformation can deliver the enormous cuts needed to limit greenhouse gas emissions by 2030: 45% compared with projections based on policies currently in place to get on track to 1.5°C and 30% for 2°C.
EGR 2023 – Broken Record	• Implementing unconditional Nationally Determined Contributions (NDCs) made under the Paris Agreement would put the world on track for limiting temperature rise to 2.9°C above pre-industrial levels this century. Fully implementing conditional NDCs would lower this to 2.5°C. • Predicted 2030 greenhouse gas emissions still must fall by 28% for the Paris Agreement 2°C pathway and 42% for the 1.5°C pathway.
EGR 2024 – No more hot air	• To limit warming to below 2°C, the annual emissions in 2030 must be 14 $GtCO_{2eq}$ lower than the current unconditional NDCs imply. • To limit global warming to 1.5°C, the annual emissions in 2030 needs to be 22 $GtCO_{2eq}$ lower. • If the 2030 emissions are not reduced, it will become impossible to set a pathway for limiting global warming to 1.5°C, and highly unlikely to achieve the 2°C limit.

Source: [16].

5.6.3.2 *Independent Assessment by WMO*

On 5 June 2024, the WMO issued a press release to the effect that

> There is an 80 percent likelihood that the annual average global temperature will temporarily exceed 1.5°C above pre-industrial levels for at least one of the next five years [17].

The statement also said that there is a 47% likelihood that the global temperature averaged over the entire five-year period 2023–2028 will exceed 1.5°C.

Can we interpret these statements into concluding that the Earth has already breached the 1.5°C limit specified in the Paris Agreement? The short answer is no. The calculation of the lower limit of the Paris Agreement temperature goal is actually based on a long-term (usually decadal) average. According to its *State of the Climate 2024 – Update for COP29* report, the WMO has this year chosen to use the latest IPCC Report (AR6) definition of global warming levels, which is in terms of 20-year averages relative to 1850–1900 levels [18]. It also shows three ways of calculating the long-term (decadal) temperature average and reveals that all three give a value of 1.3°C for 2024.

Regardless of these finer interpretations, the message delivered by the WMO can be safely interpreted as saying that the Earth is currently nearer to breaching the lower Paris Agreement temperature limit than ever before.

5.7 COP29 AND BEYOND

The last COP, held at Baku, Azerbaijan in November 2024, was held in an atmosphere of uncertainty, with the prospects of a key Party (USA) withdrawing from the climate negotiations once again looming large. While there was at least one notable success at the meeting, the general sentiment was one of extreme disappointment with the final outcomes. These are briefly summarised below.

5.7.1 Key Outcomes of COP29

The backdrop to the organisation of this meeting reveals a disturbing development in the process of climate action decision-making. According to Carbon Brief, the outcomes of COP29 had been influenced by political manipulations from the very outset of the meeting, with the Azerbaijani presidency achieved only after Russia vetoed any EU member in Eastern Europe taking up the presidency [19].

US President-elect Donald Trump's presidential victory was noted by several delegates, with varying responses. In particular [19]

- US Climate Envoy John Podesta held a press conference to reassure delegates that President Biden's outgoing team would continue to play their roles at the talks,

- EU, UK and China expressed willingness to step into larger climate leadership roles, and

- There were calls for "strengthened multilateralism".

But it was clear that Trump's election victory had a significant impact on the spirit of the climate negotiations.

The three key items of interest at the meeting were finance, fossil fuels and Article 6 Rulebook. However, of these, only the last succeeded in receiving unanimous agreement. There was bitter disappointment amongst the developing countries about finance, and discussions on fossil fuels were left in a state of limbo [19, 20].

5.7.1.1 Finance

A new climate finance goal was established at the COP29 meeting. Called the "New Collective Quantified Goal on Climate Finance (NCQG)", it replaces the earlier $100billion/yr. goal for financial assistance that the 24 developed Parties were to provide to the developing Parties.

The developing countries were campaigning for at least $1.3 trillion assistance (from the developed countries) for their mitigation and adaptation needs. The meeting agreed on $300 billion only, resulting in widespread disappointment amongst the developing Parties. According to a statement, the $1.3 trillion sum will eventually be made available by 2035 under the NCQG scheme, which will derive funding from developing countries, the private sector as well as other sources ("all actors").

5.7.1.2 Fossil Fuels

At COP28 in the UAE last year, the agreement to "transition away from fossil fuels" was encapsulated within the "global stocktake" agreements as part of the "UAE dialogue on implementing … global stocktake outcomes". At COP29, there was strong interest in this COP28 deal, and every expectation that this would be a part of the meeting agenda.

However, this item of crucial interest was to fall victim to the "agenda fight" at the start of the meeting over which items were to be included in the COP29 agenda. There was fierce debate on the inclusion of the stocktake in the agenda, but without success. Attempts were then made to include it in the discussions on the Mitigation Work Programme (MWP), but this was ultimately disallowed.

5.7.1.3 *Article 6*

As mentioned in Section 5.5 on the Paris Rulebook, agreement had not been reached at COP24 (Katowice, 2018) on the rules for internationally transferred mitigation outcomes (ITMOs) and carbon credits in the Article 6 Rulebook. These were finally dealt with at COP29. In particular

- Article 6.2 was elaborated by establishing the rules for international carbon trading (ITMOs).

- Article 6.4 was fully operationalised by laying down the rules for a new international carbon market, called the *Paris Agreement Crediting Mechanism (PACM)*. This finally provides a viable mechanism for international carbon trading on a fully commercial basis.

5.7.2 Beyond COP29

COP29 was finally successful in reaching agreement to the final details of the Paris Rulebook. It is anticipated that this will lead to a measurable movement in global carbon markets and carbon trading. Some improvement in the volume of climate funding available to the developing world has also occurred, though the overall response amongst the developing Parties to this change was one of disappointment.

In stark contrast to these small achievements, no progress at all was made on the mitigation front. This is depressing when one considers that the mitigation outcomes of COP28 had made little progress over those of COP26. This reveals that climate mitigation negotiations have virtually stalled since the beginning of this decade. It confirms that actions so far to mitigate climate change have been ineffectual and re-asserts the need for the new thinking referred to earlier in Chapters 1 and 2. This will be the subject of Chapter 9.

5.8 SUMMARY

1. Global climate policy is managed by the *UN Framework Convention on Climate Change (UNFCCC)*, a multilateral treaty that was established by the UN in 1992 amongst its members, with the aim of

stabilising human-made atmospheric GHG emissions by reducing their annual global emissions.

2. Members of the UNFCCC are called *Parties* (currently 198 in number), who meet at annual conferences called *COPs* which are organised by the *Secretariat*.

3. At first, only developed (Annex 1) Parties were required to reduce emissions. The *Kyoto Protocol*, adopted at COP3 in Kyoto, Japan, committed Annex 1 members to emissions reductions of 5.2% per annum below 1990 levels over the 2008–2012 period.

   COP13 held in Bali in 2007 adopted the *Bali Roadmap and Action Plan*, which committed developing country Parties to emissions reductions as well, assisted through technology and financial support from the developing nations.

   At COP21 in Paris in 2021, the Paris Agreement finally provided a comprehensive climate action plan, which required all Parties to reduce emissions with a view to limiting global warming to 2°C and making efforts to keep it within 1.5°C.

4. The key aims of the Paris Agreement are to hold global warming below 2°C and 1.5°C if possible by reducing emissions through Nationally Determined Contributions (NDCs) to emissions reductions by all Parties, according to the principle of "common but differentiated responsibilities".

   Voluntary contributions to NDCs are allowed (through Article 6 of the Agreement) through the establishment of Carbon Markets/Credits schemes where emissions of one country may be exchanged/traded with another.

5. The Paris Agreement did not provide details of the processes and implementation plans for the objectives stated in its Articles. These rules, known collectively as the *Paris Rulebook*, have been developed since COP24 (Katowice, 2018). The Paris Rulebook was finally completed at COP29 in Baku.

6. The success of the COPs in attaining their goals may be gauged by analysing their outcomes and assessing their efficacies via independent observers, two of which are the *UNEP* and the *WMO*.

   UNEP's progressive assessments reveal that the emissions gap (the difference between what the emissions are expected to be in 2030 and what they should be according to the Paris Agreement) has been widening, and drastic measures are needed to close it.

The latest WMO assessment of global warming, made in 2024, revealed that short-term warming averages would surpass 1.5°C within the next 5 years, and the world's long-term warming temperature average had reached 1.3°C in that year.

7. COP29 held in Baku in 2024 was successful in completing the Paris Rulebook needed to operationalise the Articles of the Agreement. It made some (though insufficient) progress on climate financing, but no progress whatsoever on the "transitioning away from fossil fuels" that was agreed to in COP28.

REFERENCES

1. United Nations Framework Convention on Climate Change. Available from https://unfccc.int/resource/docs/convkp/conveng.pdf. Accessed 22 Dec 2024.
2. Climate Change: The IPCC 1990 and 1992 Reports. Available from https://www.ipcc.ch/report/climate-change-the-ipcc-1990-and-1992-assessments/. Accessed 23 Dec 2024).
3. Sustainability for All. Achievements of the conference of the parties. Available from https://www.activesustainability.com/climate-change/achievements-of-the-conference-of-the-parties/?_adin=02021864894. Accessed 22 Dec 2024.
4. Skidmore, C., & Farrell, W. COP-Out? A brief history of the United Nations Climate Change Conferences: COPs 1–26. Available from https://www.hks.harvard.edu/sites/default/files/centers/mrcbg/programs/senior.fellows/2021-22/Chris%20Skidmore%20COP-Out%20A%20Brief%20History%20of%20the%20UN%20COP%20process.pdf. Accessed 14 Jan 2025.
5. The Bali Action Plan: Key Issues in the Climate Negotiations. Summary for policy makers. Available from http://content-ext.undp.org/aplaws_assets/2512309/2512309.pdf. Accessed 12 Jan 2025.
6. United Nations. Bali roadmap intro. Available from https://unfccc.int/process/conferences/the-big-picture/milestones/bali-road-map. Accessed 22 Dec 2024.
7. UNFCCC. Paris Agreement. Available from https://unfccc.int/sites/default/files/english_paris_agreement.pdf. Accessed 14 Jan 2025.
8. United Nations Climate Action. All about the NDCs. Available from https://www.un.org/en/climatechange/all-about-ndcs. Accessed 28 Dec 2024.
9. UNDP Climate Promise. What are NDCs and how do they drive climate action? Available from https://climatepromise.undp.org/news-and-stories/NDCs-nationally-determined-contributions-climate-change-what-you-need-to-know. Accessed 3 Jan 2025.
10. United Nations Climate Change. NDC registry. Available from https://unfccc.int/NDCREG. Accessed 28 Dec 2024.

11. Mani, F. 2020. Estimating greenhouse gas emissions in the Pacific Island Countries. In A. Singh (ed.), Translating the Paris Agreement into Action in the Pacific. Springer Nature, Switzerland AG.

12. USEPA. Inventory of US greenhouse gas emissions and sinks. Available from https://www.epa.gov/ghgemissions/inventory-us-greenhouse-gas-emissions-and-sinks. Accessed 29 Dec 2024.

13. World Resources Institute. Explaining the Paris Agreement Rulebook. Available from https://files.wri.org/d8/s3fs-public/2022-10/unpacking-paris-rulebook-english.pdf?_gl=1*m8m0us*_gcl_au*NjI4NjYzOTcxLjE3M zM5MDQ1Mzg. Accessed 30 Dec 2024.

14. UN Climate Change Conference UK 2021. COP26 goals – What did the UK Presidency aim to achieve at COP26? Available from https://webarchive. nationalarchives.gov.uk/ukgwa/20230311034236/https://ukcop26.org/ cop26-goals/. Accessed 10 June 2024.

15. Singh, A. 2023. Bioenergy for power generation, transportation and climate change and climate change mitigation. IOP Science. Bristol, UK. Chap 13. Available from https://iopscience.iop.org/book/mono/978-0-7503-3555-3. Accessed 23 June 2024.

16. UN Environment Programme. Emissions gap report 2024. 24 October 2024. Available at https://www.unep.org/resources/emissions-gap-report-2024. Accessed 1 Jan 2025.

17. WMO. Global temperature is likely to exceed 1.5°C above pre-industrial level temporarily in next 5 years. 5 June 2024. Available from https://wmo. int/files/wmo-global-annual-decadal-climate-update-2025-2029. Accessed 5 Nov 2025.

18. WMO. State of the climate 2024 – Update for COP29. Available from https://wmo.int/publication-series/state-of-climate-2024-update-cop29. Accessed 3 Jan 2025.

19. CarbonBrief. COP29: Key outcomes agreed at the UN climate talks in Baku. Available at https://www.carbonbrief.org/cop29-key-outcomes-agreed-at-the-un-climate-talks-in-baku/?utm_source=cbnewsletter&utm_ medium=email&utm_term=2024-11-24&utm_campaign=Just±publishe d±Carbon±Brief±s±in-depth±analysis±of±the±key±COP29±outcomes. Accessed 21 Dec 2024.

20. The Conversation. From a US$300 billion climate finance deal to global carbon trading, here's what was – and wasn't – achieved at the COP29 climate talks. Available from https://theconversation.com/from-a-us-300-billion-climate-finance-deal-to-global-carbon-trading-heres-what-was-and-wasnt-achieved-at-the-cop29-climate-talks-243697?utm_medium=email&utm_ campaign=Science%20newsletter%2087&utm_content=Science% 20newsletter%2087±CID_9d2ef6c5a7f66a5afc32b42d404eed55&utm_ source=campaign_monitor&utm_term=where%20were%20at%20with% 20global%20climate%20action. Accessed 14 Jan 2025.

CHAPTER 6

Strategies for Net-Zero by 2050

6.1 INTRODUCTION

The last chapter described the roles of global climate change policies and Nationally Determined Contributions (NDCs) in addressing climate change. This chapter introduces other emissions reduction schemes advocated by the Agreement. It firstly introduces strategies that take into account development-related emissions. Known as *Long-Term Low Emissions Development Strategies (LT-LEDS)*, they relate to the economic development of a country. The better-known economy-wide *net-zero strategies* are then described.

The chapter begins with an introduction to the three emissions reduction schemes, showing how they are related to the Paris Agreement and to each other and the status of their implementation by the Parties. Case studies of the application of these schemes by the UK, Australia and Fiji are then presented.

The last section deals with the framing of net-zero strategies. It is found that the net-zero strategies generally vary widely in their scope and vision, as well as in the nature and detail of their contents. It is asserted that a standardised approach is needed for the development of these strategies, and Section 6.4 proposes a development life cycle for the purpose.

DOI: 10.1201/9781003531180-6

6.2 NDCs, LT-LEDS AND NET-ZERO STRATEGIES

We saw in the last chapter that an NDC is the action plan of a Party to achieve its share of emissions reductions. In addition to NDCs, the Paris Agreement also requires Parties to reduce emissions by

- Ensuring emissions from any long-term developments that take place are kept as low as possible, and

- Establishing economy-wide net-zero strategies to achieve net-zero emissions by 2050.

All countries and developing countries in particular need to develop their economies to improve the economic wellbeing of their communities. Such developments are generally long-term in nature and invariably add to the existing emissions. The Agreement requires all Parties to implement *Long-Term Low Emissions Strategies (LT-LEDS)* that ensure that the emissions from such developments are low.

The net emission of a country is the difference between the annual carbon emissions from carbon sources and removals by carbon sinks. A net-zero strategy is a long-term, economy-wide strategy to reduce the net emissions of a country to zero in compliance with the goals of the Agreement as stated in Article 4.

How these emissions reduction schemes are related and apportioned to the Parties by the Agreement are specified in paragraphs 2, 4 and 19 of Article 4. Paragraph 2 makes it clear that all Parties are required to prepare NDCs [1]. Similarly, paragraph 19 stipulates that all Parties must also prepare and communicate LT-LEDS. The respective roles of developed and developing Parties in these emissions reduction strategies are specified in Article 4, paragraph 4, which states that

> Developed country Parties should continue taking the lead by undertaking economy-wide absolute emission reduction targets. Developing country Parties should continue enhancing their mitigation efforts, and are encouraged to move over time towards economy-wide emission reduction or limitation targets in the light of different national circumstances.

This makes it clear that the main burden of emissions reductions falls on the developed Parties, who should commit to comprehensive programmes of emissions reductions across all sectors of their economies (i.e. commit to net-zero strategies). Developing Parties on the other hand are encouraged to do what they can, according to their capabilities. From these observations, one may make the following conclusions:

- All Parties are required to prepare and communicate NDCs and LT-LEDS.

- Developed Parties are required to prepare and communicate net-zero strategies as well.

Note that the Agreement does not place any restrictions on whether a developing country can also prepare and communicate a net-zero strategy. The actual status of compliance to these obligations by the Parties was as follows at the end of 2024:

NDCs:
All 198 countries had communicated the latest NDCs [2].

LT-LEDS:
According to the UN Climate Change portal, there were a total of 76 LT-LEDS submitted by the end of the year 2024 [3].

Net-Zero targets:
Results compiled by *Statistica* show that, as of 2024, 147 out of 198 countries worldwide had some level of net-zero targets. Of this total, 5 had already achieved them, 29 had targets that were established in law, 52 had produced policy documents, and 46 countries had made proposals for a net-zero target. The remaining 51 countries had not presented a net-zero target at all [4]. These results are presented in tabular form in Table 6.1.

TABLE 6.1 Status of Achievement (as of 2024) of Net-Zero Targets by the 198 Members of the UNFCCC

State of Achievement of Net-Zero Targets by Countries (2024)					
Achieved	In Law	In policy document	Pledged	Proposed	No net-zero target
5	29	52	15	46	51

Source: [4].

6.3 CASE STUDIES

The above section introduced the three schemes used to implement the goals of the Paris Agreement and indicated how they are related. A better insight into the nature and function of these schemes is obtained by looking at actual examples, and this section examines the cases of the UK, Australia and Fiji.

6.3.1 UK

UK's main contribution to the climate mitigation effort consists of an NDC and a net-zero strategy. The NDC, submitted in December 2020, committed the UK to

> reducing economy-wide greenhouse gas emissions by at least 68% by 2030, compared to 1990 levels.

UK submitted an updated NDC in September 2022, consisting of a preamble and a filled template providing the Information to facilitate Clarity, Transparency and Understanding (ICTU), which shows how the country had strengthened the NDC submitted earlier [5].

The UK's *net-zero targets* are set via the *Climate Change Act 2008 (Amended)* [6]. This is prescribed in terms of *carbon budgets*, which set five-yearly caps on carbon emissions between the years 2008–2050 [7]. The reductions are below the 1990 levels, and the caps on five-year emissions are progressively suppressed till the target of 100% reduction below 1990 levels is reached. Table 6.2 provides the details of these carbon five-year carbon caps.

TABLE 6.2   UK's Carbon Budget, Showing How Its Five-Year Carbon Emissions Are Reduced Progressively below the 1990 levels

	Period	Carbon Emissions Cap (MtCO2e)	% Reduction on 1990 level
1	200–2012	3018	26
2	2013–2017	2782	32
3	2018–2022	2544	38
4	2023–2027	1950	52
5	2028–2032	1725	58
6	2033–2037	965	78
7	2038–2042	To be set in 2025	

Source: [7].

UK's net-zero strategy is provided in two documents: *Net Zero Strategy-Build Back Greener* (2021), produced during the Boris Johnson government [8] and *Powering up Britain: Net zero growth plan* (2023), published under the Sunak government [9]. The main feature of the strategy is the *Ten-Point Plan* for a Green Industrialized Revolution, launched in November 2020 by the Johnson government to kick-start UK's net-zero strategy. Its mission was to [8]

> create the conditions for the private sector to invest (in green technology) with confidence.

Box 6.1 shows how each of these ten points will be implemented according to the plan.

BOX 6.1   HIGHLIGHTS OF UK's TEN-POINT PLAN FOR A NET-ZERO STRATEGY

The Ten-Point Plan endeavours to achieve the aims of UK's net-zero strategy by

- Introducing legislations to ensure industry for the future demand of green products
- Mobilising GBP 26 billion of government's investment to leverage GBP 90 billion of private sector investment by 2030.

The text of column one of the Ten-Point Plan as contained in [8] is essentially reproduced in the matrix below.

	Point	Implementation
1.	Advancing offshore wind energy	• 40 GW of offshore wind energy by 2030 including 1 GW of floating wind turbines • GBP160 million into modern ports and manufacturing infrastructure • Offshore transmission network review
2.	Driving growth of low-carbon hydrogen	• Plans for 5 GW of low-carbon hydrogen capacity by 2030 • GBP240 million net-zero hydrogen fund • Trials for use of hydrogen in heating
3.	New and advanced nuclear power	• Pursuing large-scale nuclear projects, subject to value for money • Legislations for new financing model for nuclear projects • GBP385 million Advanced Nuclear Fund to enable up to GBP215 million for Small Modular Reactors

(Continued)

	Point	Implementation
4.	Accelerating the shift to zero-emission vehicles	• End of sale of new pure petrol and diesel cars and vans by 2030 and consultations on phase-out of diesel HGVs • GBP 1 billion to support electrification of UK vehicles and their supply chains • GBP 1.3 billion to accelerate roll-out of charging infrastructure • Publish a Green Paper in 2021 on the UK's post-EU emissions regulations
5.	Green public transport, cycling and walking	• GBP 120 million to begin introducing at least 4,000 zero-emission buses • Billions of pounds in enhancements and renewals of the rail network • GBP 5 billion to support buses, cycling and walking
6.	Jet Zero and Green Ships	• A Jet Zero Council • GBP15 million to support production of Sustainable Aviation Fuels • GBP20 million for the Clean Maritime Demonstration programme
7.	Greener buildings	• Plans to install 600,000 heat pumps per year by 2028 • Energy efficiency funding, including Public Sector Decarbonisation Scheme and Social Housing Decarbonising Fund • Strengthened energy efficiency requirements for private sector landlords
8.	Investing in carbon capture, usage and storage (CCUS)	Commitment for two industrial clusters by mid-2020s, and an aim for four sites by 2030 GBP 1 billion CCUS Infrastructure Fund
9.	Protecting our natural environment	• GBP 5.2 billion for flood and coastal defences • New national parks and areas of outstanding natural beauty • GBP40 million second round for the Green Recovery projects over the next four years • Establish 10 long-term Landscape Recovery projects over the next four years.
10.	Green finance and innovation	• GBP 1 billion Net Zero Innovation Portfolio (NZIP), including GBP100 million for Direct Air Capture and other greenhouse gas removal (GGR) technologies • UK's first Sovereign Green Bond • Green Jobs Taskforce

The terms and conditions of UK's net-zero strategy as described above were relaxed in September 2023 by (the then Prime Minister) Rishi Sunak in a speech which announced a revised approach with the "intention to ease the burdens on working people" [7]. A particular change was the delaying of the ban on petrol and diesel cars by five years to 2035 [10].

Keir Starmer's Labour government which followed the Conservative government in 2024 announced further changes to the net-zero strategy through the

- Great British Energy bill (to establish a publically owned clean power company)
- Crown Estate Bill (to allow for more investment in public infrastructure)
- Sustainable Aviation Fuel Bill (to facilitate the production of the fuel) [7].

6.3.2 Australia

Like the UK, Australia has also communicated both an NDC and a net-zero strategy to the Secretariat.

Australia's NDC communication follows a similar format to that of the UK. The current communication, Australia's NDC Communication 2022 [11], is an update on the country's 2020 communication. In this update, Australia announced its revised ambition for its 2030 target, which has been increased to 43% below 2005 levels. This is an improvement over the target of 23%–26% below 2005 levels set by the previous government. It also announced a

substantial and rigorous suite of new policies across the economy to drive the transition to net zero.

The details of the update were provided through a completed ICTU template.

The first substantive communication by Australia of its net-zero strategy was during the Morrison Coalition government in November 2021 [12]. It had a long-term whole of the economy plan, based around a *Technology Investment Roadmap*, to achieve net-zero by 2050. It aimed to

drive down the cost of low emissions technology" and "deploy these technologies at scale.

TABLE 6.3   The Morrison Coalition Government's Net-Zero strategy (2021)

The Morrison Government Net-Zero Strategy (2021)

Baseline year 2005	Emission reductions completed by 2020	Technology investment roadmap	Global technology trends	Use of carbon offsets	New technology breakthroughs
100%	20%	40%	15%	10%	15%

Source: [12].

The strategy used 2005 as the baseline year, noted that some 20% reductions had already been achieved and used a mix of innovative devices, global technology trends and breakthroughs to complement the Technology Investment Roadmap. Table 6.3 describes the rationale used.

Compared to the former government, the current (Albanese Labour) government of Australia has a higher reduction target and a more assured and accountable plan of emissions reductions, especially those from fossil fuels. The details are given in Box 6.2.

BOX 6.2   AUSTRALIA'S ALBANESE LABOR GOVERNMENT'S NET-ZERO STRATEGY

The Labor government has an emissions reduction target of 43% by 2030, and an ambition of making the grid 82% renewable by the same year.

The strategy to achieve this consists of a Climate Change Act [13] and a detailed programme of actions known as the *Powering Australia Plan* [14].

The Powering Australia Plan consists of

- The Safeguard Mechanism for reducing fossil fuel emissions from the grid through a capping mechanism that takes the emissions on a trajectory to net-zero by 2050.
- National Electric Vehicle Strategy that aims to improve the uptake of electric vehicles by supporting the Australian electric vehicle market.
- A Rewiring the Nation programme which is a $20 billion commitment to enhancing the national grid to enable greater penetration of renewable energy.
- A Powering the Regions fund to support industry to decarbonise and develop new clean industries, and purchase carbon credit units (ACCUs) on behalf of the Commonwealth.
- A National Reconstruction Fund of $15 billion to diversify and transform Australia's economy and industry.

Perhaps the most important feature of the Powering Australia Plan is the *Safeguard Mechanism* [15]. It assures that the emissions from Australia's

215 largest emitters are reduced by introducing a capping mechanism that eventually brings their emissions in line with the government's 2050 net-zero emissions target.

It was first introduced by the former Abbot government of Australia and required the 215 facilities to keep their emissions below a limit (or baseline) of 100,000 tonnes of $CO_{2\ eq}$. per annum. This limit, however, was too high, allowing most emitters to emit freely without the need for any reductions. The Act was amended by the Albanese government to introduce more effective baselines. Called the Safeguard Mechanism (crediting)/ Amendment Act 2023 [16], this was passed on 31 March 2023.

The amended Act reduces the emission baselines gradually by a factor of 4.9% annually till they are on a path to the 43% reductions by 2030 and net-zero emissions by 2050 trajectories. The government's Clean Energy Regulator is charged with monitoring the reductions.

The implementation of the Albanese government's strategy has had to contend with several challenges. Part of the issue evidently stems from the ambitious nature of its 43% reductions and 82% renewables targets. An example is the push-back from the rural communities of the roll-out of renewables into regional Australia, which prompted the commissioning of the Dyer Report [17, 18].

Notwithstanding these challenges, the government has been pushing ahead with its plans and has committed a sizeable proportion of its 2024–2025 budget to the cause via its *Future Made in Australia* initiative [19].

6.3.3 Fiji

Fiji is an example of a developing Party, which has lesser commitment obligations than its developed partners. In particular, it is not required to commit to an economy-wide net-zero strategy.

The country has both an NDC (updated 2020) [20] and a LT-LEDS communicated in 2018 [21]. In addition, Fiji has a Climate Change Act that was passed in 2021 [22, 23].

Fiji's NDC makes a commitment to

- achieve net-zero emissions by 2030, with 2013 as the base year

- reduce 30% of the Business As Usual (BAU) emissions (10% unconditional and 20% conditional) from the energy sector by 2030

- reach close to 100% renewable energy power generation (grid connected) by 2030

- reduce 10% BAU emissions through economy-wide energy efficiency improvements.

Fiji's LT-LEDS 2018–2030 is an economy-wide decarbonisation plan, which will

> operationalize its decarbonisation ambitions by implementing its NDC Investment Plan that is targeted towards creating a strong business case for mobilizing new and additional climate finance [20].

Note: the conditional NDCs communicated by developing countries are subject to the availability of funding, usually from the developed Parties. The mobilising of finance refers to the acquisition of such funding.

Fiji's *Climate Change Act 2021* [23] is a comprehensive and detailed undertaking to address the climate issue, starting from a declaration of climate change as a national and global emergency. The document itself comprises 17 parts, 112 sections and 2 schedules. It lists 17 objectives and goes on to establish a comprehensive response to climate change through providing for the regulation and governance of the national response to climate change, introducing a system for the measurement, reporting and verification of GHG emissions and for related matters.

Section 6 recognises and declares that Fiji and the Earth are facing a climate emergency. The document establishes the

- National Climate Change Coordination Committee,

- National Adaptation Plan Steering Committee,

- Fijian Adaptation Registry,

- Fijian Taskforce on the Relocation and Displacement of Communities Vulnerable to the Impacts of Climate Change, and

- National Ocean Policy Steering Committee.

The act charges the government to develop and implement a *National Climate Change Policy* to 2030 and for successive periods of 10 years, as well as a detailed *Transport Decarbonisation Implementation Strategy* and a *National Ocean Policy.*

It further:

- aims to provide for the relocation of at-risk communities and safeguard their rights,

- sets a net-zero emissions target for 2050,

- sets the legal framework to enable carbon sequestration, carbon stocks and emissions reduction projects,

- charges relevant agencies to conduct risk assessments and to decide on new buildings and infrastructure approvals based on resilience estimates, and

- makes provisions for the implementation of sustainable financing [23].

6.4 CASE FOR FORMAL DEVELOPMENT OF NET-ZERO STRATEGIES

The above section presented examples of net-zero strategies communicated by developed Parties to the Secretariat. It is obvious from these case studies that a wide range of methodologies can be used in the formulation of such strategies. However, it is not obvious that the methods used by a Party will ensure the achievement of net-zero emissions by the Party itself, and/or that strategies developed individually by Parties will collectively achieve the global net-zero emissions by 2050. This strongly suggests a need for a standardised approach to the development of net-zero strategies that ensures (with a high degree of confidence) that global net-zero is indeed achieved by 2050.

Motivation for such a formal approach also comes from analysing the shortcomings of the (evidently) ad-hoc approaches taken in the development of many current net-zero strategies. In particular, it is instructive to note that such strategies

- Are often limited in scope and vision – i.e. have not noted all the requirements the strategy must satisfy, or have not considered all eventualities and circumstances that may influence the viability of a strategy,

- May be non-aligned with the aims and objectives of other government strategies, leading to conflicting aims (i.e. there is a need for coherence between the net zero strategy and other strategies being developed by other government sectors)

- Are susceptible to political influence (e.g. political priorities may override the objectives of the strategy – a notable example is the shifting position of the US government with regards to its membership to the Paris Agreement).

Such a formal approach must adopt a problem-solving methodology that is tried and tested.

One such method is the well-established and understood *Systems Development Life Cycle* Approach. Such life cycles consist of several *Stages* (or *Phases*) that are followed sequentially in the development process. The advantages of this methodology are that it

- is based on empirical data and uses well-known and tested methodologies,
- has several stages that lead seamlessly from one to another, and
- provides for testing and improvement.

This is a versatile approach that

- includes all relevant aspects of the problem (i.e. determines the scope of the strategy at the outset),
- depends on measured data (i.e. gathered information, and not on opinion or speculation),
- establishes the criteria for a successful outcome at the outset,
- determines the performance of the strategy progressively, and
- uses the gathered data to make improvements.

A proposed development life cycle for net-zero strategies may be modelled after those used in information technology (see e.g. [24]).

Figure 6.1 is a flow diagram that takes the development from the initial (*Definition*) stage of the cycle, through subsequent stages of *Requirements Analysis, Design, Development, Implementation* and *Review* back to the definitions stage.

Table 6.4 describes the function of each stage of the development process depicted in Figure 6.1.

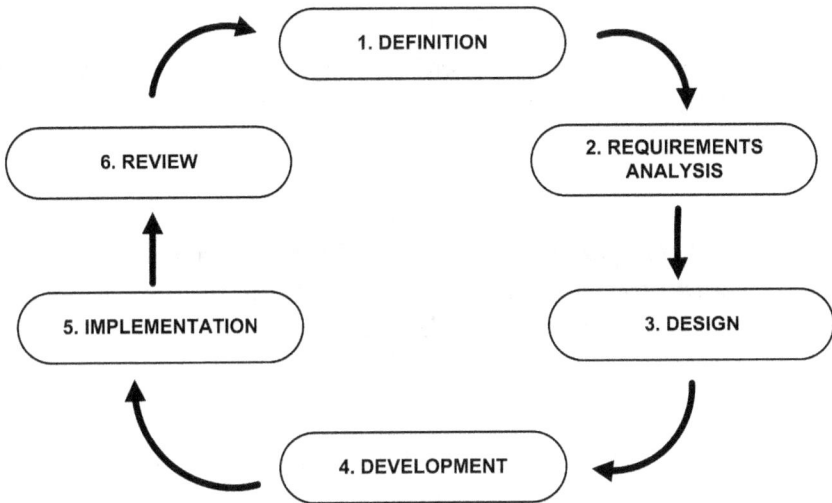

FIGURE 6.1   The flow diagram for a proposed development life cycle for net-zero strategies.

TABLE 6.4   The Proposed Stages (Phases) of a Net-Zero Development Life Cycle

Life Cycle Stage	Description
1. Definition	Define the objectives and scope of the strategy; What is the timeline for its development?
2. Requirements analysis	Gathering data to identify what is needed for a viable net-zero strategy. What are the essential and desirable criteria for a viable net-zero strategy? What are the resource requirements? What are the opportunities and challenges?
3. Design (Logical Design)	Production of a blueprint for the strategy, i.e. a conceptual description, without mentioning the specific methodologies in detail. Example: the main methods for reducing emissions are (i) replacing fossil fuels with clean energy alternatives (ii) energy efficiency (i.e. use less energy for the same task) (iii) energy conservation (avoid the use of energy by living more scrupulously).
4. Development (Physical design)	Developing the strategy itself by including the specific methodologies. Examples of methods are the carbon budget, elements of the Ten-Point Plan of the UK strategy, and the Safeguard Mechanism of the Albanese Labour government strategy.
5. Implementation	Putting the strategy into effect, e.g. passing the legislations and providing the means for initiating the strategy.
6. Review	Monitoring and evaluation of the strategy in use, and making improvements.

Note that in Table 6.4, the possibility of including differing sets of methodologies at the Design (also called the *Logical Design*) and Development (also called *Physical Design*) stages allows the development of a wide range of strategies. However, as all such methodologies must satisfy the essential criteria stipulated in Stage 2, it follows that not all will necessarily provide viable net-zero strategies.

6.5 SUMMARY

1. Three mechanisms by which emissions reductions are facilitated by the Paris Agreement are NDCs, in which countries set national targets for their emissions reductions, LT-LEDS, which ensure that long-term developments are low emissions, and net-zero strategies, which are economy-wide plans of Parties to meet their net-zero emissions by 2050 goals.

   All Parties are required to set NDC targets and communicate LT-LEDS strategies. All developed Parties are required to develop and communicate net-zero strategies.

2. The UK has communicated an NDC and a net-zero strategy. UK's Climate Change Act 2008 (Amended) sets a target of 100% reductions below 1990 levels by 2050 through a Carbon Budget scheme that caps emissions to progressively lower levels over time.

   The main feature of UK's net-zero strategy is the Ten-Point Plan, which details how reductions will be achieved through technology, energy efficiency and carbon removal methods such as off-shore wind energy, greener buildings and Carbon Capture, Usage and Storage (CCUS), respectively.

3. Australia has a new NDC target of 43% reduction below 2005 levels by 2030, which is more ambitious than the former (Morrison Coalition) government's 23%–26% reductions by the same year.

   The Coalition government's strategy was based around a Technology Investment Roadmap which drove down the cost of low emissions technologies. The new (Albanese Labour) government's strategy consists of the Climate Change Act 2022, which reflects the higher ambition, as well as a suite of actions known as the Powering Australia Plan, which consists of the Safeguard Mechanism for power production, the National EV Strategy and several funding/investment legislations.

4. As a developing nation, Fiji does not need to communicate an economy-wide net-zero strategy. It has communicated an NDC and an LT-LEDS towards its climate obligations. The most important feature of Fiji's emissions reduction initiatives is its Climate Change Act 2021, which is a comprehensive and detailed undertaking to address the climate issue. It starts by declaring a Fijian and global climate emergency and sets up the relevant administrative infrastructures to reduce the country's emissions as well as to address the impacts of climate change on the island nation.

5. It is noted in the last section that there are shortcomings in many net-zero strategies, and it is argued that a formal and standardised method should be adopted for the development of net-zero strategies. A methodology, based on a Systems Development Life Cycle Approach, is suggested which begins by establishing the key criteria for the development of viable net-zero strategies.

REFERENCES

1. UNFCCC. Paris Agreement. Available from https://unfccc.int/sites/default/files/english_paris_agreement.pdf. Accessed 5 Feb 2025.
2. UN Climate Change. NDC registry. Available from https://unfccc.int/NDCREG. Accessed 5 Feb 2025.
3. UN Climate Change. Long-term strategies portal. Available from https://unfccc.int/process/the-paris-agreement/long-term-strategies. Accessed 24 Jan 2025.
4. Statistica. Number of countries with net emissions targets worldwide, as of 2024. Available from https://www.statista.com/statistics/1549970/number-of-countries-with-net-zero-targets-by-status/. Accessed 23 Jan 2025.
5. United Kingdom of Great Britain and Northern Ireland's Nationally Determined Contribution. Updated September 2022. Available from https://unfccc.int/sites/default/files/NDC/2022-09/UK%20NDC%20ICTU%202022.pdf. Accessed 26 Jan 2025.
6. Legislation.gov.uk. The Climate Change Act 2008 (2050 Target Amendment) order 2019. Available from https://www.legislation.gov.uk/uksi/2019/1056/contents/made. Accessed 26 Jan 2025.
7. House of Commons Library. Research briefing 26 September 2024: The UK's plans and progress to reach net zero by 2050. Available from https://researchbriefings.files.parliament.uk/documents/CBP-9888/CBP-9888.pdf. Accessed 26 Jan 2025.
8. Department for Energy Security and Net Zero Policy Paper. Net zero strategy: Build Back Greener. Available from https://www.gov.uk/government/publications/net-zero-strategy. Accessed 26 Jan 2025.
9. Department of Energy Security and Net Zero. Policy paper. Powering up Britain: Net zero growth plan. Updated 4 April 2023. Available from https://www.gov.uk/government/publications/powering-up-britain/powering-up-britain-net-zero-growth-plan. Accessed 26 Jan 2025.
10. BBC News. Rishi Sunak pushes back ban on new petrol and diesel cars to 2035. 20 September 2023. Available from https://www.bbc.com/news/live/uk-66863110. Accessed 27 Jan 2025.
11. Australian Government. Australia's nationally determined contribution, communication 2022. Available from https://unfccc.int/sites/default/files/NDC/2022-06/Australias%20NDC%20June%202022%20Update%20%283%29.pdf. Accessed 29 Jan 2025.

12. Australia's Long-Term Emissions Reduction Plan. Australian government department of climate change, energy, the environment and water. Available from https://www.dcceew.gov.au/climate-change/publications/australias-long-term-emissions-reduction-plan. Accessed 28 Jan 2025.

13. Australian Government Federal Register of Legislation. Climate change Act 2022. 2022. Available from https://www.legislation.gov.au/C2022A00037/latest/text. Accessed 30 Jan 2025.

14. Anthony Albanese PM Media Releases Friday 3 December 2021. Powering Australia – Labor's plan to create jobs, cut power bills and reduce emissions by boosting renewable energy. Available from https://anthonyalbanese.com.au/media-centre/powering-australia-plan-create-jobs-reduce-emissons-renewable-energy. Accessed 30 Jan 2025.

15. Australian Government. Department of climate change, energy, environment and water. Safeguard Mechanism. Available from https://www.dcceew.gov.au/climate-change/emissions-reporting/national-greenhouse-energy-reporting-scheme/safeguard-mechanism. Accessed 30 Jan 2025.

16. Federal Register of Legislations. Safeguard Mechanism (Crediting) Amendment Act 2023. 2023. Available from https://www.legislation.gov.au/Details/C2023A00014. Accessed 30 Jan 2025.

17. ABC News. Governments urged to take control of renewable rollout to combat community distrust. Avail from https://www.abc.net.au/news/2024-02-02/distrust-anxiety-in-regional-communities-over-renewables/103419062. Accessed 30 Jan 2025.

18. Community Engagement Review. Report to the minister for climate change and energy. Available from https://www.dcceew.gov.au/sites/default/files/documents/community-engagement-review-report-minister-climate-change-energy.pdf. Accessed 30 Jan 2025.

19. Australian Government. Future made in Australia. Available from https://futuremadeinaustralia.gov.au/. Accessed 30 Jan 2025.

20. Fiji's Updated Nationally Determined Contribution. Available from https://unfccc.int/sites/default/files/NDC/2022-06/Republic%20of%20Fiji%27s%20Updated%20NDC%2020201.pdf. Accessed 1 Feb 2025.

21. Fiji Low Emission Development Strategy 2018–2020. Avail from https://unfccc.int/sites/default/files/resource/Fiji_Low%20Emission%20Development%20%20Strategy%202018%20-%202050.pdf. Accessed 1 Feb 2025.

22. Fiji Climate Change Portal. 23 Sep 2021. Available from https://fijiclimatechangeportal.gov.fj/legislation/climate-change-act-2021-an-act-to-establish-a-comprehensive-response-to-climate-change-to-provide-for-the-regulation-and-governance-of-the-national-response-to-climate-change-to-introduce-a-system/. Accessed 1 Feb 2025.

23. Climate Change Laws of the World. Fiji. Climate Change Act 2021. Available from https://climate-laws.org/document/climate-change-act-2021_8bf7. Accessed 1 Feb 2025.

24. GeeksforGeeks. Systems development life cycle. 3 Oct 2024. Available from https://www.geeksforgeeks.org/system-development-life-cycle/. Accessed 4 Feb 2025.

CHAPTER 7

Renewable Energy and the Energy Transition

7.1 INTRODUCTION

The last chapter saw examples of strategies employed by Parties to the Paris Agreement towards achieving their goals for net-zero emissions by 2050.

Four ways in which net-zero emissions can be achieved are by

- reducing the emissions from power production, transportation and manufacturing (which together account for the bulk of the emissions) as well as agriculture

- using energy more efficiently, i.e. improve energy efficiency

- reducing the use of energy, i.e. practicing energy conservation, and

- increasing the amount of carbon sinks in the biosphere through devices such as afforestation, *Carbon Capture, Use and Sequestration (CCUS), Bioenergy and Carbon Capture and Storage (BECCS)*, and *Direct Air Carbon Capture and Sequestration (DACCS)* [1].

The largest source of emissions is fossil fuel energy, which accounts for some 73% of all emissions. The most favoured way of reducing this emission is through the replacement of the fossil fuel by renewable energy (RE), which is generally less emitting than the fossil fuel counterparts. Such a transition in energy use has come to be known as the *energy transition*.

DOI: 10.1201/9781003531180-7

This chapter elaborates on the role of RE as the alternative technology of choice in the energy transition. It begins with a brief introduction to RE and its various forms. As the production of RE and renewable energy technologies (RETs) entails energy use in the various stages of their production life cycles, there are always some emissions associated with all RE and RETs. The determination of the extent of emissions associated with RE and the associated technologies requires a full *life cycle analysis (LCA)* of their production and use. This is briefly discussed in Section 7.3.

For various geographical or historical reasons, the net-zero strategies of countries differ in the type and relative importance of the RE they use in the energy mix for their energy transition [2]. Section 7.4 provides two case studies of such choices of energy mix deployed by countries, as well as examining the net-zero pathways proposed by independent energy agencies.

Nuclear energy has been proposed by several (developed) countries as an additional clean energy source. The strengths and weaknesses of such a proposal are discussed in Section 7.5. The final section raises the possibility of achieving the 1.5 and 2.0°C temperature goals of the Paris Agreement through means that do not depend exclusively on limiting GHG emissions.

7.2 A BRIEF INTRODUCTION TO RENEWABLE ENERGY (RE)

Before discussing the role of RE in the energy transition, it is instructive to consider its nature and merits as an alternative to fossil fuel energy.

7.2.1 A Natural Choice

RE is a form of energy that "never" runs out, i.e. is available from natural sources as long as these sources are available. In contrast, fossil fuels are available in underground sources in finite quantities only and are destined to be depleted in due course.

RE is also clean energy. This means that it (generally) does not emit as much carbon dioxide and other pollutants to the atmosphere as fossil fuels do. For these two reasons, RE has been (until recently) the natural choice as an alternative to fossil fuels ([2] Chapter 1).

7.2.2 Forms of Renewable Energy

RE comes in several forms, comprising solar energy, wind energy, hydro energy, numerous forms of solid, liquid and gaseous bioenergy, geothermal energy, and several types of ocean energy.

Solar energy is energy from the sun and is used for

- direct conversion into electricity through solar photo-voltaic (PV) conversion
- thermal generation of electricity via *Concentrated Solar Power (CSP)* systems
- space and water heating through use of flat plate and parabolic solar thermal collectors, and
- space and water cooling through the employment of absorption, dessicant and solar mechanical cycles ([2] Chapter 3).

Wind energy is energy derived from the kinetic energy of wind. It is used to generate electricity through *wind turbines* that extract part of the kinetic energy of wind streams and convert the extracted portion into electricity.

The maximum efficiency of a wind turbine is 59% and depends strongly on factors such as the speed and quality of the incident wind as well as the nature of the terrain within which the turbine is located ([2] Chapter 3). Wind turbines may be used either as *stand-alone* systems (i.e. isolated turbines) or arranged into large onshore or offshore arrays of turbines that form *wind energy farms*.

Hydro-energy is the (potential and kinetic) energy of water, which may be converted into electricity at hydropower plants. In the most familiar case, the potential energy of water stored in dams is used to drive turbine-electrical generator systems to produce electricity. In such cases, the power generated depends on the height difference (the *head*) between the top of the water in the reservoir and the turbine, and the flow rate of the water-jet incident on the turbine ([2] chapter 2).

Hydropower stations exist in a large range of sizes, from large plants (such as the *Three Gorges Hydro* in Sandouping, China) producing Gigawatts of power, to tiny *Pico or Nano-hydro systems* that generate less than a kilowatt of power from small rural streams for the needs of remote rural households.

A variation of hydropower technology is *Pumped Hydropower*, which is a hydro station consisting of a reservoir of water located at a significant height above the turbine, and has the capability of pumping water from the bottom of the plant back to the reservoir, preferably using other RE power sources such as wind or solar. These may form part of energy storage systems for grid energy storage.

Bioenergy is the energy stored chemically in organic matter through the process of *photosynthesis*. This process uses energy from sunlight to convert carbon dioxide and water obtained from the atmosphere and soil respectively to produce (energy-rich) *glucose* in the leaves of plants. The process is quite complex and employs *metabolic pathways* (i.e. chains of organic reactions helped by enzymes) to chemically convert carbon dioxide and water and molecules such as ADP, NADP and inorganic phosphates already existing in the plant material to produce intermediate compounds. These eventually result in the production of the (energy-rich) glucose and oxygen and the re-generation of the ADP and the NADP.

Energy captured from sunlight is stored in the form of the chemical energy of glucose, which is subsequently transferred to other organic molecules via similar metabolic pathways. The energy-rich material that is of interest to us is the *ligno-cellulosic compounds* found in the cell walls of plant cells ([2], Chapter 2).

Solid plant matter (i.e. solid biomass) can be used directly as a fuel source for combustion or converted to *secondary fuels* in the form of solid, liquid or gaseous biomass through *thermal, thermochemical, chemical, biochemical* or *electrochemical processes* ([2], Chapter 3). The forms of solid biomass feedstock used in these processes include

- dedicated energy crops,
- forestry and industrial residues,
- municipal solid waste,
- animal waste, and
- solid and liquid sewage.

Geothermal energy is energy derived from underground thermal energy sources. Such sources are usually *deep underground sources* in the Earth's crust but may also occur as *shallow underground sources* a few kilometres deep.

Geothermal energy can be used for both electricity generation and *district heating (GeoDH)*. The efficiency of geothermal power generation depends on the availability of a significant thermal gradient. The source of heat for such power generation originates from the Earth's crust which is at a temperature of ~5500°C. This maintains the Earth's mantle above it at temperatures close to 1000°C. With an average depth of the Earth's crust at about 30 km, a temperature gradient of 33°C/km near the Earth's surface (where the power station is located) is achieved. This is insufficient for a viable geothermal power plant, which requires a gradient in excess of ~40°C/km. There are, however, volcanic regions where the gradient exceeds the minimum limit. In particular, the region around the Pacific known as the "Pacific ring of fire" provides such favourable conditions.

In 2021, geothermal energy provided 14.67 GW, amounting to 0.18% of the total global generation capacity of 8012 GW. Most of this (87%) went towards heating and cooling applications, with only 13% used for electricity generation. The major heating application of geothermal energy is in *Geothermal District Heating (GeoDH)*, which employs *Ground-Sourced Heat Pumps (GSHP)* for the heating of districts, including buildings and health/recreational facilities. These use the ground at shallow depths of 2–3 km as heat reservoirs ([2], Chapter 4).

Ocean energy is the least significant of all RE technologies, contributing a meagre 0.536 GW, (or only 0.0067% of the global generation capacity) in 2021. The range of ocean technologies currently available is

- wave energy,
- tidal range,
- tidal stream,
- ocean current,
- ocean thermal energy conversion (OTEC), and
- ocean salinity gradient.

For more details on these ocean technologies, the reader is referred to ([2], Chapter 4). It must be noted that 90% of this ocean technology is due to tidal range energy, in particular the 240 MW *La Rance tidal barrage power generation* facility in France, and the 254 MW *Sihwa tidal barrage facility* in South Korea.

7.3 HOW CLEAN IS CLEAN ENERGY?

A clean energy source is ideally zero-emitting, i.e. emits zero carbon per unit energy it produces. But there is always some carbon emitted in the full life cycle chain of an energy source. As our interest in RE is to reduce net emissions by using it as an alternative to fossil fuels, it is sufficient to require that the RE fuel is less emitting than the fossil fuel it is replacing, i.e. that the *avoided emission* is positive and significant. To obtain such information requires a full life cycle analysis (LCA) of the emissions of an RE source per unit energy, and its comparison with similar emissions from the fossil fuel.

There are several processes (called *unit processes*) involved in the production and use of RE or fossil fuel. Taken together, these processes comprise the *product system* of the product (the fuel produced). Each unit process has several *inputs, outputs* and *intermediate products* that are introduced to the next unit process [3]. As an example, the production and use of the biodiesel known as *pongamia methyl ester (PME)*, obtained from the seeds of the pongamia plant, involves the processes of

- cultivating the pongamia plantation from which the pongamia seeds are derived

- extracting pongamia oil from the pongamia seeds

- converting the oil to PME through the chemical process of *trans-esterification*

- transportation of the intermediate products from one location to another

- using the PME as fuel in a *diesel engine.*

The product system for the production and use of pongamia biodiesel is shown in Table 7.1.

TABLE 7.1   The LCA Product System for PME, Showing the Unit Processes, Together with Their Inputs, Outputs and Intermediate Products

Unit Process	Inputs	Outputs	Intermediate Product
Pongamia cultivation	Diesel fuel, electricity, fertiliser, poly bags	Fuelwood, GHGs (CO_2, CH_4, N_2O)	Pongamia seeds
Oil extraction	Electricity, diesel	Seed cake, GHGs	Crude pongamia oil (CPO)
Trans-esterification	Methanol, NaOH, electricity, diesel	Glycerol, GHGs	Pongamia biodiesel (PME)
Transportation	Diesel fuel	GHGs	-
Engine combustion	PME	GHGs	-

Source: After: [3].

The total emissions per unit energy output for the production and use of pongamia biodiesel are evaluated by summing up the GHG outputs in Table 7.1. To evaluate the avoided emission, a similar product system may be constructed for the fossil fuel-derived diesel to evaluate its GHG emission, or a value for this quantity obtained from the literature. The difference found between the GHG emissions of the two production systems yields the avoided emission.

7.4 TRANSITIONING TO NET ZERO

The above two sections reveal that a very wide range of RE sources are available as alternatives to fossil fuels and discuss how the suitability of a source can be ascertained through a life cycle assessment of its emissions. This section shows how RE is being utilised in the existing and proposed transitions of the net-zero strategies of countries, starting from a consideration of the key requirements for the transitions.

7.4.1 Key Criteria for the Energy Transition

Three practical criteria for the use of RE in the energy transition are

   i. the availability of RE sources,

   ii. technologies for its conversion to the end-use form, and

   iii. conversion of intermittent sources of RE to dispatchable sources.

For a particular RE, the first requirement is the availability of the relevant resource. Such availability cannot always be taken for granted. Thus, while wind and solar energy are viable RE sources in nearly all countries, geographical and geological considerations restrict the availability of hydro and geothermal sources in many.

Energy technologies for the conversion of the various known RE sources are generally developed and market-ready. However, their actual acquisition by a country depends on such factors as the state of its development and technological advancement. This can be a serious constraint for developing nations, which often need to import the relevant RETs. In addition, the lack of supporting infrastructure and human resources for obtaining/using RETs can become a serious deterrent to their use.

Some RE sources are *intermittent*, i.e. their energy supply is not constant in time, but depends on the (generally unpredictable) state of the natural environment. In case of grid energy supply, a necessary requirement for an energy source is that it should be *dispatchable*, i.e. meet the (time varying) consumer demand at any instant in time. Thus, the energy received from intermittent sources needs to be converted to a dispatchable form before it becomes ready for use. The next section shows one method of achieving this.

7.4.2 From Intermittent to Dispatchable

A large proportion of the available RE is required for the generation of grid power. This varies with demand over the day, and sources for the grid must accommodate this daily variation in consumer demand. This is ideally achieved through the use of dispatchable sources.

Thus, methods must be found for converting intermittent sources such as wind and solar to dispatchable power. One way the issue can be resolved is through the process of *energy storage*, where the energy arriving intermittently from wind and solar energy sources is stored in an energy storage device as an interim measure, and the power for the grid is derived from the energy storage device. This combination of intermittent source and energy storage acts effectively as a dispatchable source.

Popular forms of energy storage that are available currently are battery storage and pumped hydro. The former uses rechargeable batteries (such as the popular lithium-ion battery), while the latter involves the pumping of water from a low reservoir to a high reservoir during off-peak hours.

7.4.3 Available RE Sources and Technologies

A wide range of RE sources and technologies are available for the power generation, transportation and manufacturing sectors of a country. A comprehensive treatment of such energy transition opportunities can be found at ([2], Chapters 11 and 12).

In essence, these opportunities include:

- hydro, wind and solar for power generation,

- biogas and biomethane, solid biomass and waste-to-energy, liquid biofuels, (bioethanol, biodiesel, HVO, syngas and pyrolysis oil) as fuels for power generation,

- pumped hydro, battery storage and regenerative hydrogen fuel cells for energy storage in power generation,

- rechargeable batteries and hydrogen fuel cells for EVs, liquid and gaseous biofuels for combustion engine vehicles for transportation, and

- green hydrogen for green metals and cement for the manufacturing sector.

How this range has actually been utilised in the net-zero strategies of countries is exemplified by the case studies below.

7.4.4 Case Studies of RE Use in the Transition

The net-zero strategies for the UK and Australia have been summarised in Chapter 6 and include references to the RE sources and technologies used to implement the required energy transition for each country. A fuller list of these sources and technologies is presented in Box 7.1.

BOX 7.1   RE AND RE TECHNOLOGY (RET) DEPLOYMENT IN THE UK AND AUSTRALIAN NET-ZERO STRATEGIES

UK

UK's Ten-Point Plan [4] refers to the energy sources and technologies for the energy transition required for its net-zero strategy.

The RE sources that are stated explicitly or inferred consist of

- Off-shore wind
- Green fuels such as biofuels (including sustainable aviation fuel and low-carbon hydrogen)

The RETs comprise

- Zero-emission vehicles
- Green public transportation
- Green ships
- Carbon capture, use and storage technologies, including
  - DACCS and
  - other greenhouse gas removal (GGR) technologies.

AUSTRALIA

Australia's net-zero strategy commits the nation to a target of 83% renewables by 2030 [5].

The RE sources for the energy transition include

- Wind
- Solar
- hydro (including pumped hydro) and
- green hydrogen.

The favoured RE technologies are

- Battery EVs (BEVs), hydrogen fuel cell EVs (HFCEVs) for transportation,
- Wind, solar, hydro, battery storage and pumped hydro for grid power, and
- Green hydrogen for industry (green metals and cement).

Comparison of the range of available RE sources and technologies with those actually used by the two case study nations reveals a huge gap between the wide range of available RE and those selected for the net-zero strategies of the UK and Australia, who both restrict new developments to solar, wind, hydro and use of energy storage in transportation and grid energy storage. A large range of biofuels seems to be totally ignored by both countries.

7.4.5 Proposed Global Energy Transition Pathways

The above case studies are constrained by political priorities and other limitations of the two governments and thus do not constitute impartial assessments of possible transition pathways for a country. More impartial assessments may be those produced by independent global energy agencies. Two such proposals are those proposed by the *International Energy Agency (IEA)* [6] and the *International Renewable Energy Agency (IRENA)* [7]. A brief overview of IEA's Net Zero by 2050 Scenario (NZE) is shown in Box 7.2.

BOX 7.2   IEA'S NET ZERO BY 2050 SCENARIO (NZE)

The NZE [6] is a scenario that shows what is needed for the global sector to achieve net zero by 2050. It was developed using a modelling approach that used the *Wholesale Electricity Market simulation model (WEM)* together with the *Exchange-Traded Product (ETP)* financial tool coupled with the *Greenhouse Gas – Air Pollution Interactions and Synergies (GAINS)* model.

It starts with the observation that even if all pledges made by the Parties (called the Announced Pledges Case) were met in full by 2020, global energy and industry-related emissions would still amount to 22 Gt CO_{2eq} by 2050, leading to a warming of 2.1°C by 2100. To achieve the stated goal of net-zero by 2050, the NZE advocates a strategy in which wind and solar are scaled up and traditional biomass drastically reduced over the decade (2020–2030), with new technologies developed and used over the remaining two decades to achieve the desired result by 2050.

In the NZE, by 2030:

- Global energy and industry-related emissions fall by 40%, and methane emissions from fuels fall by 75%,
- Solar and wind become the main sources of electricity generation,
- Half of the emissions reductions are facilitated through the use of energy efficiency, solar and wind, and
- The use of traditional biomass is phased out.

Beyond 2030:

- New emissions reductions are provided by increased electrification of the end-use sector and newer technologies such as hydrogen and *carbon capture, use and sequestration (CCUS).*
- Behavioural changes amongst people and businesses also contribute significantly to emissions reductions.

The above goals are achieved through the evolution of the global total energy supply (TES; see figure 2.5 of [6]) along a pathway that is characterised by a combination of the electrification of end use energy sector, energy efficiency, behavioural change in people and businesses and a drastic reduction of traditional bioenergy in favour of modern bioenergy. Fossil fuels fall from 80% of TES in 2020 to 20% in 2050.

In the resulting global emissions pathway (see figure 2.2 of [6])

- Global energy and industry-related emissions fall to 21 Gt CO_{2eq} in 2030 and net-zero by 2050
- Advanced economies achieve net-zero by 2045, and they collectively remove 0.2 Gt CO_{2eq} from the atmosphere by 2050 via CCUS (including Direct Air Capture and Carbon Storage (DACCS) and Bioenergy and Carbon Capture and Sequestration (BECCS).

TABLE 7.2   Comparison of the Estimated Total Energy Mix in 2020 and 2050 in the NZE. (EJ = Exajoules = 10^{18} Joules)

		2020	2050	
Total energy mix		590 EJ	550 EJ	
Individual sources	Oil	30% (~ 177 EJ)	Total fossil fuel	~ 110 EJ
	Coal	26% (~ 153 EJ)		
	Natural Gas	23% (~ 135 EJ)		
	RE	17% (~ 100 EJ)	RE	~ 390 EJ
	Nuclear	5% (~25 EJ)	Nuclear	~ 50 EJ

Source: [6].

Note that many of the values in the table above have been estimated from figure 2.6 of [6]. For a better appreciation of the comparative values, the reader is referred to the original source.

RE plays a critical role in the energy transition espoused by the NZE. As seen in figure 2.18 of [6], the RE share in power generation increases from 29% in 2020 to 60% by 2030 and 88% in 2050. While the main RE contributor to generation was (dispatchable) hydro in 2020, this is replaced by a mix of hydro (12% of generation), bioenergy (5%), concentrated solar power (2%) and geothermal (1%) in 2050.

Bioenergy plays a significant role in the NZE, providing 25% of the total global energy supply in 2050. The supply of bioenergy (consisting of traditional and modern bioenergy) was 65 EJ in 2020. Some 90% of this was traditional biomass. This form of energy is eliminated in the NZE due to its inefficiency and health concerns, and modern bioenergy increases from less than 40 EJ in 2020 to 100EJ in 2050.

The *TES* in the NZE (see figure 2.5 of [6]) sees a shift from a predominantly fossil fuel base in 2020 to largely renewables in 2050. Fossil fuels reduce to a quarter of their 2020 value in 2050, while renewables increase to four times their value in 2020. The range of fuels in the energy mix in 2050 is much more diverse than in 2020 and includes wind, solar, hydro and other RE, modern bioenergy, natural gas, oil, coal and nuclear. Table 7.2 provides a summary of the salient features of the TES in 2020 and 2050.

It must be borne in mind that IEA's NZE is the output of economic modelling tools (WEM, ETP and GAINS). A consequence of this method of development is that the final output is not unique, i.e. there are many other possible pathways to achieving net zero than the one presented, and the NZE presented is only one instantiation of many possible pathways.

7.5 HOW GOOD IS NUCLEAR ENERGY?

Some developed countries, including Australia, are considering the inclusion of nuclear energy as an additional clean energy source in their energy transition mix [8, 9]. The justification given for the choice of nuclear energy as an additional clean energy source is that it is both clean and safe. It is prudent to subject these claims to diligent analysis.

In Chapter 6, it was pointed out that the net-zero strategies of some countries evidently had certain shortcomings, indicating the need for a formal (standardised) procedure for the development of net-zero strategies in general. One deficiency of the strategies was a disregard for certain essential criteria that were key requirements of net-zero strategies as a whole.

Two such essential criteria for net-zero strategies are that they must

- Deliver substantive and verifiable reductions in emissions to ensure net-zero emissions by 2050, and

- Be aligned with the objectives of other government strategies, i.e. must be coherent with these strategies.

The suitability of nuclear energy as a candidate for the energy transition can be tested using these two criteria. The first criterion can be tested by carrying out an LCA of this energy source and its associated technologies. The claim for its "zero emission" status becomes debatable when one considers the embedded (embodied) emissions associated with nuclear fuel, as the following analysis shows.

The path from the acquisition to the final disposal of the fuel for nuclear reactors (usually uranium) consists of several processes. According to the IAEA [10], this *nuclear fuel cycle* consists of the following four stages:

- *Conversion* – where the Uranium Oxide ("Yellow Cake") derived from the uranium ore is converted to gaseous uranium hexafluoride (UF_6).

- *Fuel fabrication* – in which uranium dioxide UO_2 is made into ceramic pellets, which are sintered, milled, and packed into long metal tubes made of zirconium alloys, several of which make up fuel assemblies.

- *Power generation* – in which the nuclear fuel is used in a reactor for 3–6 years.

- *Spent fuel storage* – where the spent fuel rods are removed from the reactor and stored under water for cooling and radiation shielding and later moved to another pool or air-cooled shielded buildings for a total of 40 years. In the last phase of the cycle, the spent fuel is permanently removed from the reactor site and can be safely disposed in deep underground (geological) repositories, *the first of which were to be commissioned by Finland and Sweden in 2020.*

The entire process is heavily energy-intensive. In particular, the conversion stage (where the uranium fuel is *enriched* to increase the percentage of the fissionable ^{235}U isotope from 0.7% to 3%–5% in the normal isotopic mixture of ^{235}U and ^{238}U) consumes inordinate amounts of energy (with the release of associated CO_2 emissions). Thus, the claimed "clean" status of nuclear energy depends critically on the amount and nature of the fuel source used in the *enrichment process*, and the "clean" claim should be treated with caution.

Most UN member nations actively support the *Sustainable Development Goals (SDGs)*. Goal 3 of the SDGs is to ensure health and well-being for all at all ages. To remain coherent with the SDGs, the net-zero strategy of a country must therefore ensure that its implementation does not present health and safety risks to its communities.

In this regard, the fourth stage of the nuclear fuel cycle (spent fuel storage) is of great concern. According to the World Nuclear Waste Report [11].

Even 70 years after the beginning of the nuclear age, no country in the world has found a real solution for the radiating legacy of nuclear power. .... No country has a final disposal site for nuclear waste in operation yet; Finland is the only country that is currently constructing a permanent repository. Most countries have yet to develop and implement a functioning waste management strategy for all kinds of nuclear waste.

The permanent repository is the same as the deep underground (geological) repositories mentioned in stage 4 of the nuclear fuel cycle above. This implies that all nuclear power plants around the world (with the possible exception of Finland and Sweden) are still storing their high-level nuclear waste at surface sites, or re-processing them for re-use.

Indeed, according to IAEA's publication on the status of spent fuel management [12], about one third of all spent fuel produced between the start of the nuclear age in 1954 to 2016 has been reprocessed, while

> the remaining two thirds are stored, pending processing or disposal (in deep geological repositories).

This publication also reveals that (up to 2016)

> Most spent fuel was held at nuclear plant sites in wet storage in reactor pools.

Apart from the obvious health and safety hazards due to such storage, a new hazard to human wellbeing has emerged in the current geo-political scenario in the form of *weaponising of nuclear waste*. This presents the opportunity for unfriendly entities of attacking the exposed high-level nuclear waste with the aim of dispersing highly radioactive material to the surrounding region. Such intentions were a cause of concern in the Russian drone attacks on ZNPP [13].

That such activities pose a real threat to nuclear safety has been confirmed by statements made by the IAEA. Because of such concerns, this global nuclear watch-dog has been physically monitoring the nuclear sites in Ukraine over most of the last three years [14]. This response from the IAEA confirms that weaponing of nuclear safety is a new nuclear hazard that the world will have to contend with henceforth.

7.6 A BETTER GOAL THAN NET ZERO?

Is net-zero global emissions the only way to achieve our climate aims?

Climate change is caused by global warming, which occurs when the Earth's energy balance is upset (see Chapter 4). Therefore, the main aim of climate action should be to focus on the cause of the imbalance and to limit the warming to 1.5°C (rather than focusing solely on achieving net zero). When one adopts this approach, it is found that there are several possible ways in which global warming can be restricted, by far the most important and obvious of which is the reduction of the atmospheric GHG concentrations to a specified level. But the other possibilities are worthy of consideration as well.

There are other factors, apart from the physical ones noted above, that determine the achievement of the final goal. The implementation of climate strategies can be severely restricted by political decision-making, as was seen in the recent withdrawal of the US from the Paris Agreement and the reversal of its fossil fuel policies. Emissions reductions can also be expedited by economic policies that promote positive tipping points in the economic domain. In short, a set of actions from the scientific, economic, political and geo-political domains can be used to maximise the ultimate chances of success in climate action.

How such actions can be integrated to produce a strategy spanning several domains will be discussed in the last two chapters of the book.

7.7 SUMMARY

- RE is a natural choice as an alternative for fossil fuels, as it never runs out wherever its sources are available, and is less emitting than fossil fuels.

- The principle forms of RE sources consist of solar, wind and hydro which can be converted directly to electricity; solid bioenergy that can be converted to secondary fuels before use, as well as gaseous biofuels such as biogas and biomethane; geothermal energy which is less important, providing only 0.18% of total global generation capacity; and the least important ocean energy, comprising wave, tidal, ocean current and ocean thermal energy conversion (OTEC), which amounted to only 0.0067% of the global generation capacity in 2021.

- RE is not completely emissions-free. However, as an alternative to fossil fuel, it is sufficient that the avoided emissions it produces are positive and significant. Such information can be obtained by carrying out a LCA of the RE and comparing it with the similar analysis of the fossil fuel.

- Three essential requirements for the use of RE in the energy transition are that the relevant resources are available, the technology for the conversion of the RE for end-use are developed and market-ready, and viable mechanisms exist for the conversion of intermittent forms of RE to dispatchable energy.

- RE sources/technologies that are available for use in the power (electricity) generation, transportation, and manufacturing sectors are

  - hydro, wind, solar and a range of solid, liquid and gaseous bioenergy for power generation

- battery storage and pumped hydro for energy storage needs in power generation

- rechargeable batteries and hydrogen fuel cells for BEVs, HFCEVs, E-Scooters in transportation, and

- green hydrogen for green metals and cement for the manufacturing sector.

- The UK government's *Ten Point Plan* and Australia's net-zero strategy provide case studies of now the available RETs have been deployed in the net-zero strategies of countries. The cases reveal the enormous gap between the available RETs for the transition and the (limited) selection that has been used in practice.

- Idealised energy transition pathways that are not subject to political and budgetary constraints have been proposed by *the International Renewable Energy Agency (IRENA)* and the *International Energy Agency (IEA)*, which are independent global energy agencies. IEA's *Net Zero by 2050 Scenario (NZE)*, developed with the assistance of WEM, ETP and GAINS modelling tools, reveals a possible pathway to 2050 that begins with a total global energy mix of 590 EJ in 2020 that is dominated by fossil fuels. This evolves to a mix of 550 EJ in 2050 where RE becomes the major component.

- Claims by the Australian government opposition that nuclear energy is a clean and safe alternative for the energy transition are evaluated by comparing them with essential criteria for net-zero strategies. It is firstly found that the claimed "clean" status of nuclear energy depends critically on the amount and nature of the fuel source used in the enrichment of the nuclear fuel. Secondly, it is noted that nuclear power plants will require the disposal of high-level nuclear waste (containing spent fuel rod assemblies) generated on a regular basis. Currently, these are stored *on site* at all nuclear power plants, posing unacceptable health and safety risks in the emerging new geo-political order.

- There are other (non-physical) factors that influence the achievement of the climate goal of limiting warming to 1.5°C. These include the role of decision-making and the global geo-political status. How the success of the climate goal can be enhanced by integrating such considerations into a new global strategy is considered.

REFERENCES

1. Gambhir, A., & Tavoni, T. Direct air carbon capture and sequestration: How it works and how it could contribute to climate-change mitigation. One Earth 1(4) (Dec 2019) 405–409. Available from https://www.sciencedirect.com/science/article/pii/S2590332219302167. Accessed 3 April 2025.

2. Singh, A. March 2025. Talking Renewables (Second Edition) – A Renewable Energy Primer For Everyone. IOP Publishing. Bristol, UK. Available from https://iopscience.iop.org/book/mono/978-0-7503-6280-1. Accessed 20 Mar 2025.

3. Singh, A., & Charan D. 2020. Life cycle analysis as a tool for estimating avoided emissions. In Singh, A., & Deo, R. (eds.), Translating the Paris Agreement into Action in the Pacific. Springer Nature, Switzerland.

4. Department for Energy Security and Net Zero Policy Paper. Net zero strategy: Build Back Greener. Available from https://www.gov.uk/government/publications/net-zero-strategy. Accessed 26 Jan 2025.

5. Australian Government Federal Register of Legislation. Climate Change Act 2022. 2022. Available from https://www.legislation.gov.au/C2022A00037/latest/text. Accessed 30 Jan 2025.

6. International Energy Agency (IEA). Net zero by 2050. A roadmap for the global energy sector. May 2021. Available from https://iea.blob.core.windows.net/assets/ad0d4830-bd7e-47b6-838c-40d115733c13/NetZeroby2050-ARoadmapfortheGlobalEnergySector.pdf Accessed 16 Sep 2024.

7. IRENA. World Energy Transitions Outlook 2023. 1.5°C Pathway. Vols. 1 and 2. 2023. Available from https://www.irena.org/-/media/Files/IRENA/Agency/Publication/2023/Jun/IRENA_World_energy_transitions_outlook_2023.pdf. Accessed 16 Sep 2024.

8. Dutton, Littleproud, O'Brien Media Release – Australia's Energy Future. 1 June 2024. Accessed 27 Aug 2024.

9. ABC News. Peter Dutton reveals seven sites for proposed nuclear power plants. Accessed 27 Aug 2024.

10. IAEA. Getting to the Core of the Nuclear Fuel Cycle. Department of Nuclear Energy. International Atomic Energy Agency,. Vienna, Austria. Available at https://www.iaea.org/sites/default/files/18/10/nuclearfuelcycle.pdf. Accessed 3 April 2025.

11. World Nuclear Waste Report. The World Nuclear Waste Report 2019. Focus Europe. Available from https://worldnuclearwastereport.org/. Accessed 4 April 2025.

12. IAEA Nuclear Energy Series. Status and trends in spent fuel and radioactive waste management. IAEA. Available from https://www-pub.iaea.org/MTCD/Publications/PDF/PUB1963_web.pdf. Accessed 4 April 2025.

13. CNN News. Russian-controlled Zaporizhzhia nuclear reactor damaged following drone attack. 8 April 2024. Available from https://edition.cnn.com/2024/04/07/europe/russian-controlled-zaporizhzhia-nuclear-reactor-damaged-following-drone-attack/index.html. Accessed 4 April 2025.

14. IAEA. Update 271. IAEA Director General Statement on the Situation in Ukraine. 23 Jan 2024. Available from https://www.iaea.org/newscenter/pressreleases/update-271-iaea-director-general-statement-on-situation-in-ukraine. Accessed 4 April 2025.

CHAPTER 8

Beyond 1.5°C

Taking Stock

8.1 INTRODUCTION

It was noted in Chapter 2 that the current methods of addressing climate change were not working (see Section 2.4 for reasons of failure) and that there was a need for new thinking to deal with this seemingly intractable problem. In particular, a way had to be found for re-framing the climate problem that widened the scope of the solution set and provided insights into new possibilities.

This is the intention of the final two chapters of the book. This chapter looks at the current status of global warming and the most imminent *climate tipping points (CTPs)* predicted by climate models and contemplates the possible scenario that will result in the new climate equilibrium beyond 1.5°C. The next and final chapter of the book proposes a new climate action framework that incorporates climate action from other domains to produce a new, all-inclusive strategy for achieving the climate goal.

The next section of this chapter takes a more detailed look at the CTPs first considered in Chapter 4. Our special concern is to see what threat tipping points might present to living organisms. To assess the severity of their impacts on life forms, it is first necessary to understand what the requirements for life are under normal circumstances. This is elaborated in Section 8.3. In the following two sections, a model to visualise CTPs is adopted, followed by a detailed look at possible climate scenarios in the

DOI: 10.1201/9781003531180-8

new climate equilibrium after 1.5°C. Section 8.6 describes observed cases of climate impacts that could indicate the nature of those that would follow after 1.5°C, and the final section speculates on the nature of the future climate scenario.

8.2 CLIMATE TIPPING POINTS NEAR 1.5°C

The key aim of this chapter is to investigate what the world's climate will be like as predicted by global warming and the several CTPs that are imminent around a warming temperature of 1.5°C. According to WMO, this temperature limit may be reached as early as June 2030 (see Chapter 2). As the health and welfare of humans and other living forms are of the utmost importance, the main focus is to identify the threats that such conditions will pose to the survival and wellbeing of all living forms and the ecosystems to which they belong.

The threats can be better understood by comparing them to the basic requirements and optimal conditions for the survival of life. To achieve this aim, one needs to

   i. List and critically examine the possible CTPs near 1.5°C,

   ii. Enumerate the optimal conditions for life on earth,

   iii. Examine the possible scenarios in the new climate regime near 1.5°C, and

   iv. Compare these with the enumerated optimal conditions to evaluate the threats.

The list of possible CTPs over a large range of global warming temperatures was provided in Table 4.4. Our present interest is in investigating those that are likely to occur near 1.5°C, as it is likely that this Paris Agreement temperature limit might be reached within the latter part of this decade [1]. These CTPs have been elaborated by *the European Space Agency (ESA)* [2] and are presented in Table 8.1.

The achievement of the stated aim of this chapter relies on the implicit assumption that the combined effect of the CTPs can be obtained by adding together the effects of all CTPs. We note however that the CTPs are generally located in distinct geographical locations of the Earth and may occur at different times, i.e. are distributed in time and space. Our primary interest is to know what the impacts of the CTPs will be at any one location.

TABLE 8.1   Climate Tipping Points (CTPs) Near 1.5°C

CTP (with Type, Location and Temperature Threshold)	Description (After ESA [2])
1. Boreal permafrost (Abrupt thaw, North Canada, < 2°C)	Permafrost (i.e. ice-rich frozen soil) undergoes rapid thaw and ice melt, causing subsidence of land surface with formation of lakes. Reduces land and other natural resources used by the local population.
2. Boral forest southern (Die-back, Russia, < 2°C)	The southern edges of this Boreal forest either declines or dies-off, leading to change in ecosystem (and loss of biodiversity) over decades.
3. Greenland ice sheet (Collapse, Greenland, < 2°C)	Loss of ice sheet over several thousand years, with faster collapse at higher temperatures, leading to rising sea levels impacting coastal regions globally.
4. Labrador sea and sub-polar gyre (collapse, North Atlantic Ocean, < 2°C)	The collapse of this ocean surface current system (driven by wind), will cause disruptions in the regional ocean circulation pattern, influencing regional/global climate systems through connections to the Atlantic Meridional Overturning Circulation (AMOC).
5. Barent Sea winter ice (abrupt loss, Arctic Ocean, < 2°C)	Winter sea ice in the Barent Sea may be abruptly lost around 1.5°C due to inflow of warm Atlantic water. This will impact the regional ecosystems.
6. Low latitude coral reefs (die-off, Indonesia and Australia, < 2°C)	Ocean water reaches a threshold temperature where coral begins to bleach, triggering widespread coral death. Ocean acidity also contributes to this coral ecosystem collapse.
7. West Antarctic ice sheet (Collapse, Antarctica, < 2°C)	This collapse is due to the self-sustaining and rapid disintegration of the ice sheet and can lead to substantial global sea level rise impacting coastal regions.
8. Amazon Forest (dieback, South America, (2–4°C)	Deforestation of the Amazon rainforest is threatening to create a tipping point where the wet rainforest, which creates its own rainfall, may switch to a dry grassy Savannah ecosystem. This will, amongst other things, diminish the natural resources of the rainforest available to the indigenous peoples of the region.

Source: ESA [2].

A partial solution to this difficulty emerges when one realises that several of the CTPs may happen more or less concurrently. It would then be of interest to note any possible interactions between the CTPs, so that it is meaningful to sum together the net impacts. A review by Wunderling et al reveals that interactions are indeed possible, occur over several time scales and are largely destabilising [3].

In principle, therefore, the above procedure may be employed to obtain a meaningful picture of the distribution of climate conditions due to the net effect of global warming and the possible triggering of several CTPs. These conditions may be compared to the basic requirements or optimal conditions for survival of life to differentiate between those that are

i. Life-threatening (which apply to all living organisms), and

ii. Those that threaten the wellbeing of organisms, and in particular the economic wellbeing of humans.

The requirements or conditions for life are addressed in the next section.

8.3 BASIC REQUIREMENTS FOR SURVIVAL

The adverse impacts of climate change may be broadly categorised into *slow onset events* [4], *extreme weather events* and the *impacts of CTPs on the climate system*. The former two impacts are well-known and include sea-level rise, melting sea ice, land degradation, hurricanes/storms, wild fires and heat waves, cold spells, draughts, floods and more.

Global warming and the triggering of CTPs can change the environmental conditions for living forms from those that are optimal for survival to those that cause physical stress leading to possible death. Before we can evaluate the hazards that a changed climate system can pose to living forms, and humans in particular, it is essential to obtain a clear picture of the optimal conditions in which they live and thrive.

For their survival and wellbeing, all living organisms require [5–7]:

- a physical environment with provides a supply of oxygen for organisms depending on aerobic respiration for their energy needs,

- the existence of tolerable ranges of environmental variables such as temperature, pressure, and relative humidity, pH and salinity,

- adequate food and nutrition, and

- a healthy ecosystem.

Humans on their part also need:

- a socio-economic system with a stable economy,

- a system of governance,

- International relations, and

- global ethical standards.

These requirements are presented as a matrix in Table 8.2. Important highlights of the table are:

- The *environmental conditions* that living organisms can tolerate are determined by evolutionary history. One feature of this history is that life cannot exist outside the liquid phase temperature range of water, and it is found that most organisms cannot survive in conditions where the temperature falls outside the 0°C– 45°C range [8].

TABLE 8.2   Basic Requirements for the Survival and Wellbeing of Living Organisms

Requirements	Examples	Remarks
All Living Organisms		
1. Physical environment	Oxygen supply	Both plants and animals need oxygen for (aerobic) respiration.
	Tolerable temperature, pressure and relative humidity ranges	For land-based flora and fauna
	Tolerable temperature, pressure, pH and salinity ranges	For ocean-based life forms
2. Food and nutrition	Adequate food, water, sunlight as well as the biogeochemical cycles.	Food and water are essential for humans. Sunlight and the biogeochemical cycles (including the carbon, nitrogen and phosphorus cycles) ensure the availability of energy and nutrition for plants.
3. Ecosystem	Each organism is an integral part of its own ecosystem	Ecosystems make up the biosphere
Additional Requirements for Humans		
1. Socio-economic system	An economy that provides the basic goods and service needs of the population	
2. System of governance	A stable government	
3. International relations	An international trade system between nations	
4. Ethical standards	A basic (universal) morality that provides the basis, and the final motivation, for global social cohesion.	

- Biogeochemical cycles play a critical role in supplying essential elements to living organisms by converting the inorganic forms obtained from the environment to organic molecules which are part of the metabolisms of these organisms and cycling them back to the environment to complete the cycle [9, 10].

- All organisms require an aqueous medium within their cells for the metabolic processes that produce the organic molecules essential for growth and reproduction. This water is provided by the *water cycle* [9, 10].

- The biogeochemical cycles are driven by energy from the sun and transfer the elements from inorganic sources in the environment to the cells of living organisms for the production of complex organic molecules via metabolic processes, and finally return the elements to the environment through death and environmental re-assimilation

- The cycles are part of the organism's ecosystem, which is a part of the global biosystem (we are all inter-dependent!).

How will these conditions for life and wellbeing be affected by global warming and the imminent CTPs listed in Table 8.1? To find out, we need to obtain a picture of what the new climate equilibrium will be like after 1.5°C. To do this, we need a working model that describes what happens at a CTP. This is the subject of the next section.

8.4 CONCEPTUALISING THE NEW CLIMATE EQUILIBRIUM

Recall that a CTP was defined in Chapter 4 (Section 4.5.1) as the temperature threshold at which part of the Earth's climate system, called the tipping element, changed irreversibly. But how do we visualise such a CTP?

A CTP can be visualised by representing it with a physical model. The task is expedited by making the following assumptions:

- The Earth's climate system can exist in at least two equilibrium states,

- Crossing (or triggering) the CTP involves switching from the current climate equilibrium to a new equilibrium, and

- At a CTP, the Earth's climate switches from its current equilibrium state to a new climate equilibrium beyond a temperature known as the *temperature threshold.*

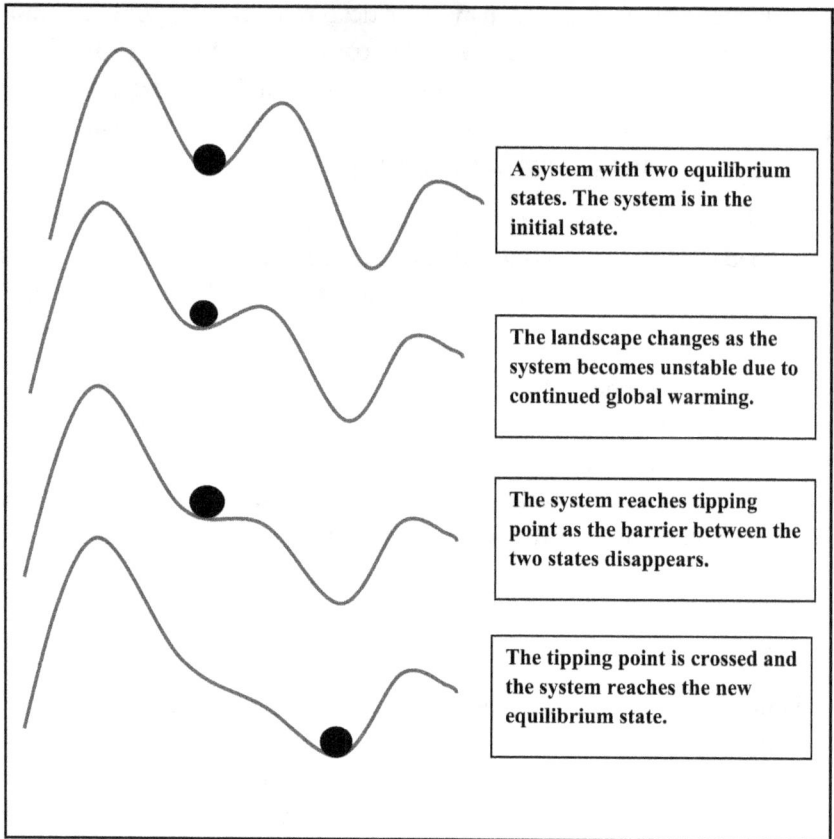

FIGURE 8.1   A physical model of a climate tipping point. (Figure adapted from ESA [2].)

A dynamic model incorporating these requirements has been produced by ESA. It was first described in words in Chapter 4 (Section 4.5.5). Figure 8.1 provides a graphical representation of the same model through four snapshots of its states.

8.5 POSSIBLE SCENARIOS IN THE NEW CLIMATE EQUILIBRIUM

Before considering the actual (observed) climate impacts beyond 1.5°C, one must first consider what climate science predicts about them. This is essentially summarised in the latest relevant reports published by the IPCC. The first such report was the Special Report on 1.5°C [11] produced in 2018, followed later by the contribution of Working Group I to the sixth assessment report (AR6 WG 1 Report) [12]. These reports established that, according to the available science at the time, there would be a sharp increase in the severity of climate impacts in going from a warming temperature limit of 1.5°C to 2.0°C and proposed schemes for categorising and assessing the severity of these impacts.

The following three sub-sections introduce the characterisation scheme, explain the significance of 1.5°C, and how it was evaluated using climate and ancillary models by the IPCC.

8.5.1 IPCC's Classification of Climate Impacts

There is a range of types of climate impacts that are of concern. The IPCC Special Report on 1.5°C characterises these impacts in terms of risk levels organised into five broad categories known as *Reasons for Concern (RFCs)* (see [11], Figure SPM.2, p. 11). Table 8.3 lists these categories and their descriptions.

The levels of risk for a severe impact for each category are ranked as *very high, high, moderate* and *undetectable* and generally increase with global warming temperature.

A new scheme for the description of climate impacts, based on *Climate Impact Drivers (CIDs)*, is introduced in the AR6 WG1 Report Contribution to the Sixth Assessment Report (AR6) [13] (see Figure SPM.9, p. 26). This report defines CIDs as physical climate system conditions (e.g. means, events, extremes) that affect elements of society or ecosystems.

The CIDs are grouped into seven types: *Heat and Cold, Wet and Dry, Wind, Snow and Ice, Other, Coastal* and *Open Ocean*. To give an example

TABLE 8.3   IPCC's Reasons for Concern (RFC) Classification of Climate Impacts

RFC Category	Examples	Remarks
1. Unique and threatened systems	Coral reefs, Arctic sea ice, ecosystems	These systems are very susceptible, even at 1.5°C
2. Extreme weather events	Heatwaves, floods, droughts	Risk rises steeply after 1.5°C
3. Distribution of impacts	Who suffers the most?	Poorer countries suffer the most, even at 1.5°C
4. Global aggregate impacts	Total economic and ecological loss	There is less loss at 1.5°C compared to (say) 2.0°C
5. Large-scale singular events	Ice sheet collapse, massive ecosystem die-offs	These are more likely beyond 1.5°C

Source: [11].

of their contents, the Heat and Cold group contains the elements *Mean Surface Temperature, Extreme Heat,* and *Cold Spells.*

For each of these CIDs, the AR6 WG1 Report Figure SPM.9 shows the number of land and coastal regions (which are reference regions of the world defined in the Atlas chapter, Figure Atlas.2) that are impacted by the respective driver.

The IPCC's AR6 projections show that

- All of the groups experience changes in at least 5 CIDs,

- 96% will experience changes in at least 10 CIDs, and

- half in at least 15 CIDs.

The reader is referred to Figure SPM.9 for predictions of the nature of the change (i.e. increase or decrease, and the number of reference regions affected).

8.5.2 The IPCC and 1.5°C

What is important about a mean global warming of 1.5°C, and how can one ascertain the state of the climate beyond it? How did the IPCC arrive at this critical temperature?

The task can be achieved through the assistance of climate modelling tools to ascertain the global mean temperatures that result from various trial emissions scenarios and the respective emissions pathways. The resulting climate impact at each such temperature can be used to identify the critical temperature beyond which the impacts become unsustainable.

The details of such a procedure are discussed in Box 8.1.

BOX 8.1   HOW MODELLING CAN BE USED TO ARRIVE AT THE 1.5°C WARMING TEMPERATURE LIMIT

A climate model uses emissions pathways relating to specific mitigation scenarios as inputs, and

- outputs the global warming temperature limit that will result
- shows when it will peak during the century
- outputs other significant climate impacts such as sea level rise and frequency of extreme weather events.

The model is run for different emissions pathways as inputs, and the warming temperature outputs are noted. The critical temperature is determined by the emissions pathway for which the severity of the climate impact outputs is considered to be just sustainable.

Mitigation pathways used by climate scientists are the Shared Socio-economic Pathways (SSPs), where

- SSP1-1.9 has very low emissions, producing a radiative forcing of 1.9 W M^{-2}
- SSP1-2.6 has low emissions, with a corresponding radiative forcing of 2.6 W M^{-2}
- SSP2-4.5 has intermediate emissions and a radiative forcing of 4.5 W M^{-2}
- SSP3-7.0 has high emissions, resulting in a radiative forcing of 7.0 W M^{-2}
- SSP5-8.5 has the highest emissions, with a radiative forcing of 8.5 W M^{-2}

Example:
Typical mitigation Pathways used as inputs are:
Mitigation pathway P1
In this pathway, concerted actions have been taken to reduce emissions, and consequently

- **CO_2 emissions** fall by about 45% by **2030** relative to **2010** levels.
- **Net-zero CO_2 emissions** are reached around **2050**.

Also,

- use of **bioenergy with carbon capture and storage (BECCS)** is kept to a minimum, and
- energy efficiency, renewable energy and lifestyle changes are maximised.

Mitigation pathway P4
This is a pathway in which little mitigation action has been taken, and

- **CO_2 emissions** peak at **2030** before beginning to decline.
- **Net-zero emissions** occur **after 2070** and depend on carbon dioxide removal and storage (CDR).

The procedure involves running pathway P1 and noting the climate system outputs, and repeating the process with pathway P4.

The outputs for P1 are:

- Global warming remains **below 1.5°C** throughout the century.
- The peak warming of about **1.5°C** occurs **around mid-century**, then stabilises.
- **No significant overshoot** of 1.5°C eventuates.
- **Sea-level rise** is expected to be about **26–77 cm** by 2100.
- **Frequency of extreme heat**: about **14% of the world's population** is expected to be exposed to severe heatwaves at least once every 5 years under 1.5°C.

The outputs for P4 are:

- **Peak warming** is expected to be **1.8°C** and occurs around **mid-century.**
- **Return to 1.5°C**: The warming reduces to 1.5°C by around **2100**, but only with the help of aggressive negative emissions.
- **Sea-level rise** is about **28–82 cm** by 2100.
- **Risk of tipping points** is high and includes possible irreversible ice sheet loss or ecosystem collapses.

As the P1 emissions pathway is one of the lowest emission mitigation pathways sustainable by economies, it should be clear from a comparison of the above results that 1.5°C is close to the limiting warming temperature that will avoid widespread severe climate impacts.

8.5.3 Modelling Specific Impacts Near 1.5°C

How can we characterise the climate impacts for a specific global warming temperature? The procedure is as follows:

Choosing coral reef bleaching as the specific example:

- Step 1: Use an Earth Systems Model (ESM) to simulate ocean temperature under different mitigation. pathways (e.g. pathways P1 and P4 above) and obtain respective temperature output.

- Step 2: Input this temperature output of the ESM into an ecological model appropriate for coral reef bleaching.

- Note the different outputs for levels of bleaching from both inputs.

The actual modelling is described in Box 8.2.

**BOX 8.2   MODELLING THE IMPACT OF CLIMATE
CHANGE ON CORAL REEFS**

A procedure to evaluate the impacts of global warming on the health of coral reefs is the following:

- Use an Earth Systems Model (ESM) (e.g. as in the CMIP5 project) to simulate the ocean temperature profiles for representative pathways (use pathways P1 and P4).
- Input the temperature output from the ESM into an *ecological model* that simulates coral reef health, such as bleaching thresholds.
- A suitable model is one which evaluates the Degree-Heating-Weeks (DHS) for corals (where 1 DHW = 1 week of sea surface temperature (SST) at 1°C above normal SST, and a value of 4-8 produces mass bleaching and mortality).
- Note the differences.

Results:
For a scenario described by emissions pathway P1, around 70%–90% coral reef loss occurs at a global warming temperature of 1.5°C.

For a P4 scenario, there is a high likelihood of more than 99% coral reef loss, occurring at a global warming temperature of 1.8°C.

For a report on a typical research carried out on coral bleaching using ESM modelling, the reader is referred to the paper by Frieler et al. [14].

8.6 OBSERVED HUMAN AND ECOSYSTEM IMPACTS OF CLIMATE CHANGE

The above three sub-sections presented the results of climate modelling similar to that performed by the IPCC. They are projections of climate impacts, i.e. are theoretical predictions. So what are the actual impacts as observed experimentally? The short case studies below provide examples of the actual climate impacts, including heat waves, coral bleaching events and forest ecosystem collapse that have been observed in the recent past.

HEAT WAVES CASE 1: THE 2003 SUMMER HEATWAVE IN EUROPE

A well-documented and analysed heatwave event was the summer heat wave in Europe in 2003, as reported by UNEP [15].

According to this UN agency, maximum recorded temperatures during this severe heatwave and drought ranged between 35 and 40°C during July and August of 2003 in the southern Europe region extending from Germany to Turkey. It killed thousands of people (mostly the elderly), caused wildfires and had severe impacts on ecosystems, agriculture and glaciers. Power and transport disruptions followed, and agricultural production declined, resulting in losses in excess of 13 billion euros.

With a total death toll estimated to be in excess of 30,000, this was one of the ten mostly deadly disasters in Europe during the last century. The deaths attributed to this event in various European countries are shown in Table 8.4.

The agricultural sectors of the affected countries were affected directly through the impacts on crop harvests and the complex inter-relationship in the natural food supply chain as follows:

- The extreme weather conditions decreased both the quantity and quality of harvests, with the extreme heat accelerating crop development by several weeks leading to early ripening.
- The combination of high temperatures and increased solar irradiance boosted crop water consumption, resulting in depletion of soil water and low crop yields.
- Some of the most impacted sectors were the green fodder supply, livestock and forestry sectors, as well as potato and wine production. The livestock farmers also suffered in the following winter due to lack of green fodder.
- The drought affected vegetation by lowering photosynthetic activity which led to a reduction in the productivity of crops as well as fodder. It also affected forests by weakening trees and making them more vulnerable to diseases and insects.

TABLE 8.4 The Recorded Death Toll in the European Summer Heatwave of 2003

Country	Deaths Attributed to the Heatwave
France	14,082
Germany	7000
Spain	4200
Italy	4000
UK	2045
Netherlands	1400
Portugal	1300
Belgium	150

Source: [15].

HEATWAVES CASE 2: THE HEATWAVE OF JULY 2023

More recently, a heat event of global proportions occurred that broke all temperature records in several locations around the northern hemisphere [16].

According to the World Weather Attribution (a scientific group devoted to attributing extreme weather events to global warming),

> July 2023 saw extreme heatwaves in several parts of the Northern Hemisphere, including the Southwest of the US and Mexico, Southern Europe and China. Temperatures exceeded 50°C on the 16th of July in Death Valley in the US as well as in Northwest China. Records were also reached in many other weather stations in China and the all-China heat record was broken in Sanbao on the 16th of July. In Europe, the hottest ever day in Catalunya was recorded and highest-ever records of daily minimum temperature were broken in other parts of Spain. In the US, parts of Nevada, Colorado and New Mexico tied their all time high ... highest ever night time temperatures in Phoenix Arizona which also had its record for longest time without falling below 90F/32.2C.

This group attributed these extreme heat events to human-induced climate change by noting, on the one hand, that

> "North America, Europe and China have experienced heatwaves increasingly frequently over the last years (~2023) as a result of warming caused by human activities, hence the current heat waves are not rare in today's climate, with an event like the current expected approximately once every 15 years in the US/Mexico region, once every 10 years in Southern Europe, and once in 5 years for China", and observing further that

> "Without human induced climate change these heat events would ... have been extremely rare. In China it would have been about a 1 in 250 year event while maximum heat like in July 2023 would have been virtually impossible to occur in the US/Mexico region and Southern Europe if humans had not warmed the planet by burning fossil fuels".

CASE 3: GLOBAL CORAL BLEACHING EVENTS (APRIL 2025)

Between 1996 and the present, the world has experienced four coral bleaching events that resulted in severe coral bleaching/coral reef loss around the world. The main cause of these un-natural occurrences is higher-than-normal ocean temperatures which turn the coral white and lifeless and have a devastating effect on the marine ecosystem that uses the reefs as its habitat. The fourth event is currently ongoing.

The Guardian [17] reports that

- the current (4th) event began in January 2023,
- 84% of reefs are being exposed to high (bleaching-level) ocean temperatures, compared to 68% during the third event lasting between 2014 to 2017, 37% in 2010 and 21% in the first event in 1998,
- at least 82 countries and territories are experiencing bleaching conditions that turn corals white,
- several areas have experienced bleaching continuously for several years, including the Great Barrier Reef,
- recently, the Ningaloo coast in Western Australia experienced its highest levels of heat stress on record, and
- bleaching has also been reported from reefs off Madagascar and the east African coast.

Coral bleaching is not always irreversible, and recovery is possible if temperatures are not too extreme. However, recent outcomes of bleaching events are a cause for alarm.

CASE 4: FOREST ECOSYSTEM COLLAPSE (APRIL 2024)

Forest ecosystem collapse becomes evident when a forest loses its normal form and begins dying out and is unable to regenerate itself. There is consequent loss of biodiversity. There are multiple causes (or stressors) for the collapse, which may be both natural and human-made. They may include (climate change-induced) increase in temperature, drastic changes in rainfall, drought, wildfires on the one hand, or human-created de-forestation and land use change on the other [18].

A recent example is the Western Australia forest ecosystem event reported by the Australian ABC news [19]. According to this report, after a six-month drought (believed to be the driest spell on record), the state was experiencing conditions similar to the forest collapse of 2010/2011. Symptoms included the browning off of the water-stressed plants, which began dying off in patches.

In an earlier study by the Australian Geographic [20], the journal found 19 Australian ecosystems that could be classified as "collapsing". The list included

- The dry interior of the continent,
- The savanna and mangrove regions of Northern Australia,
- The Great Barrier Reef,
- Shark Bay,

- South Australia's kelp and alpine forests,
- The tundra region on Macquarie Island, and
- The moss beds in Antarctica.

As an example of the sequence of multiple stressors that could contribute to a forest collapse, some 40 million mangrove trees died along a 1000 km stretch of the Gulf of Carpentaria in late 2015 due to extreme temperatures (30°C lasting for 18 days), lengthy drought conditions, and influences due to feral pigs, scrub fires and invasive weeds. Another significant contributor was the El Nino, which caused a prolonged drop in sea level of about 20 cm which deprived the mangrove trees of water.

The recovery of the mangrove forest has since been slowed by two severe cyclones and floods. Growth of seedlings has been hampered by continued "tidal rafting" of the dead tree trunks. The economic consequences of the damage to the region's A$30 million fishing industry have been significant.

The difference between modelled projections of climate impacts and actual observations of these changes cannot be emphasised too much. One is a predicted result, while the other gives examples of what is actually happening. The science projects that the observed impacts will become worse in the future, and judging from the success of the scientific predictions, this is a matter of utmost concern.

8.7 RECONCILING CTPs AND THE WORK OF THE IPCC

Before leaving the topic of the observed impacts of climate change, it is important to reflect on issues relating to the work of the IPCC and papers by specific groups on CTPs reported more recently in the literature.

Ideally, the entire climate scenario after 1.5°C should be predictable by climate science, with the projections improving with the gathered understanding. The IPCC uses the assistance of expert volunteer authors to

> assess the thousands of scientific papers published each year to provide a comprehensive summary of what is known about the drivers of climate change, its impacts and future risks....

The results are published in IPCC Assessment Reports and Special Reports [21]. Towards these ends, the IPCC collaborates in the World Climate Research Program (WCRP) with modelling centres around the world to model future climate scenarios for the Assessment Reports (see Chapter 4).

Thus, it would seem that all possible impacts that can be predicted by climate scientists should be included in the reports published by the IPCC, and that this includes the work of scientific groups working in the specific area of CTPs.

In practice, however, it must be acknowledged that the IPCC reports require extended periods of time to prepare, during which time the latest results relating to CTPs may escape the notice of the IPCC authors. One thus expects

- the IPCC reports to include references to the earlier work on CTPs, but not necessarily to the latest published papers,

- the latest work on CTPs published by specific groups working in the area may appear in the literature earlier than the IPCC reports.

The papers and reports sourced in the preparation of this book have come from both IPCC reports and more recent publications by others in the scientific literature, and the reader is advised of possible discrepancies due to the time lag outlined above.

8.8 WHAT ARE THE THREATS?

We now return to the main focus of this chapter, namely identifying the likely threats that the new climate conditions after 1.5°C will pose to the survival and wellbeing of all living forms and the ecosystems to which they belong.

A general picture can be obtained from an assessment of the modelling as well as the observed results presented in the earlier sections. Referring back to Section 8.3, the most essential and urgent requirements for the survival of living organisms are the environmental conditions, followed by food and nutrition. Any threats to the denial of these conditions to living organisms are threats to their survival and wellbeing.

To prioritise these threats, we note that *the most serious are those that threaten human life within the shortest timespans.* Apart from oxygen depletion (see later) *extreme heat stress is the most urgent of these threats. Food unavailability (famine) is the other obvious threat.* As organisms can (arguably) survive for longer periods without food or nutrition, this may be termed as a "delayed threat".

An assessment of how these threats may develop in time may be obtained by analysing the 4 case studies presented in section 8.6 above.

Specifically, cases 1 and 2 show an increase in maximum temperatures (from 35°C to 40°C to 50°C) in the regions experiencing the heatwaves of 2003 and 2023, respectively. These global extremes have been treated more comprehensively in the UN Secretary-General's Call for Action (see below).

In the case of ecosystem collapse, the coral bleaching events as exemplified by the global coral bleaching event of 2024 reveal a clear upward trend in the geographical extent of coral bleaching over time (from 21% of the total coral reefs in 1998 to 84% in 2023). The dependence of bleaching on the ocean surface temperature was clearly demonstrated in Box 8.2, where *modelling revealed 70%–90% coral loss at a warming temperature of 1.5°C and in excess of 99% at 1.8°C.*

These projected and observed results strongly suggest that a large proportion of the world's coral reefs are under a serious threat of being lost almost completely by 2030, by which time Corpenicus estimates that the 1.5°C limit would have been reached (see Chapter 2).

Death due to heat stress remains the most serious concern and indications are that the toll due to this environmental condition will increase greatly in future heatwaves.

Support for this view comes from the UN Secretary-General's call for action on 25 July 2024 [22]. The pdf copy of this call is available at [23].

According to the documentation of the Secretary-General's call,

> Extreme heat dwarfs the impact of more visible weather hazards such as tropical cyclones (16,000 deaths/year). Model estimates show that between 2000 and 2019, approximately 489,000 heat-related deaths occurred each year, with 45% of these in Asia and 36% in Europe.

and

> By 2050, if the current trend continues, every child under 18 in the world (2.2 billion) will be exposed to high heat wave frequency (up from 24% in 2020).

The difficulty in assessing the true impacts of extreme heat stress is in determining the number of deaths, which require collection of data from death certificates. These are not always easy, especially in the case of developing countries.

8.9 SPECULATING THE FUTURE

What are some of the worst possible future scenarios imaginable?

The following examples are considered concerning enough to be worthy of attention:

8.9.1 Oxygen Depletion

Is oxygen depletion possible?

In Table 8.2, adequate oxygen supply was listed as the most crucial requirement for nearly all living forms. What is the possible threat posed by climate change to this vital supply?

The Atlantic Meridional Overturning Current (AMOC) is a system of currents that circulates water in the Atlantic Ocean, bringing warm water northwards (as a surface current) and taking cold water southwards (as a deep water current) [24]. It is part of the global *thermohaline circulation*, a current that acts as a conveyor belt and joins all the currents of the world.

The surface current carries phytoplankton, which is the beginning of the food chain for all marine life. These planktons grow by capturing sunlight and obtaining their nutrient supply from the deep current when it upwells and mixes with the surface current in the South [25].

The likelihood of ocean-based oxygen depletion has been raised in relation to the predicted collapse of AMOC [26, 27]. While debate persists in the literature about the likelihood of such an event, the IPCC AR6 WG1 Report has provided unambiguous evidence that there has indeed been a loss of 0.5%–3.3% dissolved oxygen in the upper 1000 m depth of the open ocean between 1970 and 2010. It also states that

> Warming, via solubility reduction and circulation changes, mixing and respiration are considered the major drivers, with 50% of the oxygen loss for the upper 1000 m of the global oceans attributable to the solubility reduction [28].

The experimental evidence must clearly supersede any results from modelling. One must await improved modelling results and/or more substantive experimental data to obtain a clearer picture.

There are no reports so far of similar land-based oxygen depletion possibilities due to climate change. This, however, should not be reason for complacency, as final decisions will rest on the outcomes of scientific investigations that are still ongoing.

8.9.2 The Socio-Economic Reality Beyond 1.5°C

Predicting the how the socio-economic status of a region influences the impact of climate change is not an easy task because of the sheer complexity of the possible scenarios. Some insights can be gained from the Integrated Assessment Models (IAMs) that are currently being used routinely for predictions of Economic futures [29, 30].

A more direct way of expressing the differential nature of future climate impacts on societies is through the following (hypothetical but compelling) scenario, which shows how the socio-economic status within the demographic distributions of the world can be the ultimate factor that determines the severity of the climate impacts felt by humans.

Box 8.3 explains how this can happen.

BOX 8.3   SOCIO-ECONOMIC STATUS AND THE SEVERITY OF A CLIMATE IMPACT – A HYPOTHETICAL SCENARIO

The socio-economic status of a country or region may make a climate impact more severe for poorer countries than for the rich, even though the physical attributes of the climate impact may be the same. This (hypothetical) scenario demonstrates how this can happen in the case of extreme heat.

In the UN Secretary-General's Call for Action, extreme heat was called the "Silent Killer" presumably because of the difficulty in deciding when it had taken its toll. But it is also important to emphasise how this climate impact chooses to act selectively and to the detriment of the less economically advantaged on the global and national scales. The following (hypothetical) scenario is aimed at demonstrating how the poor are often less prepared to mitigate the fatal effects of extreme heat stress than those who are more economically capable.

Assuming that all humans possess the same temperature tolerance, the impact of heat stress on humans depends on its severity, which is determined by how high the ambient temperature is and the time for which the heatwave is present.

Now the effects of extreme heat stress can be mitigated through various means, the most common and available means being the reduction of the ambient temperature of the living space through cooling (e.g. air conditioning (a/c)). This requires the availability/affordability of a/c technology and the electricity to power it.

The issue arises that not all regions of the world may have the same access to air-conditioning or the money to acquire it or the electricity to power it. The poor people, in particular those residing in slums and informal settlements, will naturally be less capable of acquiring the a/c to reduce the severity of this silent threat to life.

TABLE 8.5   Proportion of Urban Population Living in Slums or Informal Settlements in Selected Countries

Country	SDG Sub-region	Proportion of Population in Slum/Informal Settlements (%)	Date of Reference	Total Population (2025)
Canada	Northern America	1.1	2022	41,300,000
France	Western Europe	0.0	2021	68,467,000
Germany	Western Europe	0.0	2009	83,456,000
Russia	Russia	2.6	2022	146,299,000
UK	Northern Europe	0.2	2022	68,500,000
US	Northen America	0.1	2022	337,014,000
Angola	Middle Africa	62.7	2022	39,040,000
Benin	Western Africa	64.0	2022	14,814,000
Congo	Middle Africa	75.3	2022	6,484,000
Argentina	South America	14.5	2022	45,851,000
Bolivia	South America	46.6	2014	12,581,000
Bangladesh	Southern Asia	51.5	2022	175,686,000
Cambodia	South East Asia	42.3	2022	17,847,000

Source: [31].

Table 8.5 shows data collected by UN-Habitat on the proportion of urban populations living in slums or informal (squatter) settlements in selected developed and developing countries of the world. It immediately reveals pronounced differences between the developed and the developing countries in the proportion of their people living in slums and informal settlements. While the developing countries have a substantive proportion of their population living in informal settlements, the proportion of informal settlers in the developed countries is generally close to zero.

For the sake of this exercise, one can assume that most of the people living in normal homes have access to a/c as a means of cooling during periods of extreme heat. Adopting a figure of 50% for slum/informal settlers indicates that *in the event of a heatwave of the same severity occurring over all regions of the world*

- *some 12 million people will be in extreme danger of dying in Angola (population 39,040,000), while*
- *only 227,000 will face similar threat in Canada (population 41,300,000), and*
- *France and Germany (with larger populations) will face minimal threat.*

A similar rationale can be applied to the fate of the populations of the *Least Developed Countries (LDCs)* (population 1,100 million in 2022).

According to *SDG 7 Energy Progress Report 2024* [32], electricity access in the LCDs in 2022 was 57%. On the assumption that access to a/c is equal to access to electricity, one finds that *some 400 million people in the LDCs will be exposed to the same threat to life* under similar heat stress conditions as for the countries mentioned in Table 8.5 above.

These estimates are disturbing and portray a possible future scenario where millions become casualties of extreme heat stress in developing countries while developed countries are left essentially unharmed. According to the current emissions estimates, increases in global warming temperatures are expected to continue till at least 2030. This paints a grim picture of our imminent climate future.

8.9.3 Total Biosphere Collapse

Ecosystems are integral parts of the biosphere. We have seen evidence of the collapse of individual ecosystems. So what is the possibility of the whole biosphere collapsing – i.e. a total biosphere collapse?

To obtain an insight into such a possibility, one first needs to know what the biosphere is, what its components are, and have a better understanding of the remarkable inter-dependency within and between these components. This is the aim of this section. The next sub-section begins with an introduction to the biosphere.

8.9.3.1 A Brief History of the Biosphere

The biosphere may be defined simply as the part of the Earth and its atmosphere comprising all living organisms and their environment. According to *Britannica*, it is essentially a global ecosystem which consists of living organisms (the biota) and the non-living (abiotic) environment that provides the living organisms with their energy and nutrients [33].

The Earth was formed some 4.5 billion years ago (BYA). The first signs of life began appearing about 1 billion years later, i.e. 3.5 BYA [34, 35]. The first organisms were the unicellular *bacteria* and *archaea*, which lived in the ocean and did not need oxygen to survive.

About 2.6 billion years ago, some of the bacteria (known as the cyanobacteria) acquired the ability for photosynthesis, a process in which solar energy is used to convert water and carbon dioxide into oxygen and energy-rich glucose molecules [34]. This event is sometimes called the "oxygen revolution". The ultimate result was that the ocean eventually became saturated with this new element, which then began entering the atmosphere. This finally enabled the origin of organisms in both the ocean and on land that used oxygen to obtain their energy needs (through the process of aerobic respiration).

The early single-celled organisms (i.e. bacteria and archaea) did not have distinct cell nuclei and other organelles that are present in the cells of multi-cellar organisms. These multi-cellular organisms originated from another group of micro-organisms known as the *eukaryotes* some 1.6 billion years ago. The first evidence of such multi-cellular organisms is from 600 million years ago (MYA).

8.9.3.2 Ecosystems

An ecosystem is a system comprising a community of living organisms and the environment on which they depend [36, 37].

Ecosystems may be conveniently divided into *biotic* and *abiotic* components. *Biotic components* comprise all the living organisms (plants (the flora), animals (the fauna) and other living forms. The *abiotic components* are non-living components and include the physical environment as well as the climatic conditions.

Ecosystems are classified as either *Terrestrial* or *Aquatic*. The 12 major ecosystems of the world are:

- the *Tropical Rainforest, Temperate Forest, Taiga* or *Boreal, Tundra, Scrubland, Desert* and *Grassland ecosystems*, which comprise the Terrestrial ecosystems, and

- the *Lentic, Littoral, Lotic, Coral reef* and the *Oceanic ecosystems*, which are the Aquatic ecosystems

Ecosystems can be grouped together into *biomes*. These are large regions of land, sea or the atmosphere comprising the ecosystems and the geography and climate [38].

The unique feature of an ecosystem is the inter-dependency between its parts. The living organisms as a whole (the biota) depend on the inorganic environment for their energy and nutrients, and the various communities of organisms rely on each other to obtain these requirements. This is made possible through the food chains the organism is part of and the biogeochemical cycles on which the organism depends on.

8.9.3.3 Food Chains and the Biogeochemical Cycles

Living organisms are made up of organic molecules, which require energy and nutrients (containing essential elements) to produce and maintain. They obtain these essentials through food chains and the BGC cycles. How this transfer of energy and nutrients is achieved is best seen by describing the flow of energy through the food chain of an ecosystem.

The flow of energy through an ecosystem is described by its *Trophic Levels*. This is the position an organism occupies in its food chain [39].

The energy flow in food chains in both the terrestrial and aquatic ecosystems consists of the following sequence:

Starting from Plants (the producers of energy), the energy flows to the

- Herbivores (who eat plants), who are eaten by the

- First Level Carnivores, who in turn are eaten by the

- Second Level Carnivores.

Each stage in this energy sequence is called a *Trophic Level*, with the plants having the highest trophic level, and second-level carnivores the lowest. The difference in the levels of energy is due to energy losses in going from the starting level (plants – the producers of energy) to the second level carnivores (who have the lowest levels of energy).

The food chains frequently occur in *food webs*, which link several food chains, possibly having the same starting and endpoints. To take a (somewhat fictitious) example, a food chain may start from vegetables and proceed to crickets, birds and humans in that order. But other alternatives that have the same beginning and end are

- Starting from vegetables and going directly to humans, and

- From vegetables, to cows, to humans.

All three taken together form a food web that provides three alternative food routes for humans to source their food needs from.

Food chains are indispensable for the survival of all living forms, as they ensure them the essential requirements of energy and building materials (in the forms of atoms and organic molecules). However, they do not show how living forms obtain their continuing supply of these nutrients, which are present in abundance in the abiotic components of the Earth's ecosystems. This connection to the non-living resources of the Earth is provided by the *biogeochemical (BGC) cycles*.

Living organisms depend on six main elements to produce most of the mass of their cells. These are

- Hydrogen (H),

- Oxygen (O),

- Carbon (C),

- Nitrogen (N),

- Phosphorous (P), and

- Sulphur (S).

In addition, the metallic elements *calcium* and *magnesium* are essential for the formation of organo-metallic complexes. Altogether, there are *sixteen different elements found in all living organisms* [40].

Each element has its own BGC cycle, which obtains an organically insert form of the element from the abiotic component of the ecosystem, converts it to an organic form (usually with the help of micro-organisms) and makes it available to plants for the production of the *macromolecules of life*. These plants are eaten by herbivores who in turn may be eaten by carnivores, all of which depend on macromolecules for their existence. The cycle is completed when the element is returned to the abiotic component of the ecosystem (usually via microbial action) through death and decay.

As an example, in (a highly simplified version of) the nitrogen cycle, nitrogen molecules present in the atmosphere are "fixed" by nitrogen-fixing bacteria in the root nodules of certain plants to form *ammonia* in the soil. Other bacteria in the soil convert this to *nitrites* and *nitrates* which can be assimilated by plants. Yet other bacteria in the soil reduce the nitrate to form *nitrogen molecules* (a process known as de-nitrification) which are returned to the atmosphere in this abiotic form.

When plants and animals die, their bodies (containing the four macromolecules of life) are decomposed by bacteria and fungi in the soil to re-produce ammonia.

Together with the help of other branches of the nitrogen cycle, nitrogen is kept in circulation between the air (where it exists as the abiotic nitrogen molecule) to microbes, plants and animals in the biota on the Earth's surface, to the soil (as ammonia, nitrites and nitrates) and back to the atmosphere via

- various inorganic forms in the soil (ammonia, nitrites, nitrates),

- organic forms (the four macromolecules of life) in the microbes, plants and animals, and

- to the atmosphere again through the process of de-nitrification.

In summary,

- *Food chains provide the essential nutrients and energy needs of all living organisms and transmit these essential requirements through the biotic component of the ecosystem.*
- *The biogeochemical (BGC) cycles obtain these essentials from the abiotic components of ecosystems, cycle them through the biotic component, and return them to the abiotic environment after the organism dies.*
- *The whole process is driven by energy from the sun, which is obtained by plants through the process of photosynthesis.*

8.9.3.4 Life and the Biogeochemical Cycles

Why are the biogeochemical (BGC) cycles so important to life?

The BGC cycles enable the formation of the (organic) macromolecules that are necessary for all life. The four macromolecules necessary for plants and animals are the

- Carbohydrates (or polysaccharides)

- Lipids

- Proteins, and

- Nucleic acids.

A macromolecule is a very large (organic) molecule that is usually made up of repeating subunits called *monomers* [41].

Carbohydrates are energy-providing molecules that occur as *monosaccharides, disaccharides* or *polysaccharides*, with the monosaccharide providing the monomers from which the larger molecular chains are made.

Lipids do not have a structure consisting of repeating units and occur as the *triglycerides, phospholipids, carotenoids, steroids* and the *waxes*. *Proteins* have *amino acids* as their monomer units. Twenty such amino acids provide the building blocks of these macromolecules. *The nucleic acids* consist of *Deoxyribonucleic acid (DNA) and Ribonucleic acid (RNA) made up from monomer units called nucleotides.*

All these molecules have carbon, hydrogen and oxygen in common and may also contain nitrogen, phosphorus and sulphur as additional atoms. Table 8.6 lists the four types of macromolecules of life and their elemental composition.

It is seen from Table 8.6 that

Without the six main elements needed for life, the four macro-molecules of life would not exist.

As the elements are supplied by the six respective biogeo-chemical (BGC) cycles, the growth and wellbeing of the biota depends critically on the existence of the cycles. It follows that any disruption to the BGC cycles is likely to have a devastating impact on the biosphere as a whole.

8.9.3.5 Signals for a Total Biosphere Collapse

If a total biosphere collapse is possible, how can we know if the collapse is imminent?

We can gain some insights into this question by asking what happens during an ecosystem collapse, and considering the links between the different ecosystems that make up the biosphere. Ecosystems occur collectively in the form of biomes, and as a first step, it may be instructive to

TABLE 8.6   The Four Types of Macromolecules of Life and Their Constituents

	Macromolecule	Monomer Units and Examples	Elemental Composition
1.	Carbohydrates (polysaccharides)	Monomer units: monosaccharides e.g. glucose, fructose, galactose. Examples: starch, cellulose.	C, H, O
2.	Lipids	Diverse group of macromolecules, including triglycerides, phospholipids, carotenoids, steroids, waxes.	C, H, O, P
3.	Proteins	Monomer units: amino acids (~ 20 in number). They contain the carboxylic (-COOH) and the amino (-NH$_2$) groups. Examples: Alanine, Aspartic acid, Cysteine, Histidine.	C, H, O, N, S
4.	Nucleic acids (DNA, RNA)	Monomer units: nucleotides. Consist of pentose sugar, phosphate and base group.	C, H, O, N, P

Source: [41].

investigate the conditions under which the collapse of one ecosystem leads to a similar fate for another.

But there are even wider linkages. Oxygen supply is a common necessity for most of the ecosystems on both land and sea. Most of the terrestrial ecosystems depend on atmospheric oxygen for their survival, while most aquatic ecosystems cannot survive without the presence of dissolved oxygen in the ocean. The need for water (as a medium for cellular metabolism) is just as universal, and the two ingredients work hand in hand to maintain life.

A total biosphere collapse may be defined as the sudden or gradual decline and eventual demise of the Earth's biota, perhaps beginning with the collapse of individual ecosystems which trigger cascades through the biota of the Earth. These events may be triggered by certain events, which, in the case of forest collapse/dieback for instance, could include impacts such as drought and heat-related forest stress [42].

A possible contributing factor to biosphere collapse is the failure of food supply chains for plants and animals, leading inevitably to the failure of human food supply chains. A significant proportion of these chains originates in food farms. According to the Food and Agriculture Organisation (FAO), the world's (traditional) smallholder farms produce around one third of the world's food (while the rest is produced by larger "modern" farms) [43]. This fraction, however, is contested by some non-governmental organisations [44], who believe the figure could be as high as 70%.

Regardless of whether it is traditional or modern, the agricultural farm is essentially a managed ecosystem. The farmer maximises the production of these ecosystems through tilling, planting and the use of fertilisers and pesticides, as well as techniques such as crop rotation and cover cropping (which are natural means of managing ecosystem health). In the possible event of future climate change-induced extreme droughts, such farms could experience collective failure on the global scale, with serious consequences on the food industry and the global food supply system.

In general, identifying the beginnings of a collapse is a difficult task. But one can begin by noting crucial *points of vulnerabilities (POVS)* within ecosystems, as well as the triggering events for observed cases of ecosystem collapse. *The BGC cycles clearly qualify as important POVs.* Food chains are closely associated with the BGCs. Oxygen (and water) availability are also factors to be concerned about. So far, there are little signs of oxygen depletion, but it is wise to remain vigilant.

The occurrence of droughts/heat waves is an obvious trigger for ecosystem collapse and a good candidate for a biosphere event. A signal for an impending collapse could be the increasing frequency of disasters such as droughts, heat waves and floods.

8.10 SUMMARY

1. Several CTPs are imminent near a global warming temperature of 1.5°C, with potential threats to the survival and wellbeing of living organisms. These CTPs are distributed in both space and time, but interactions between those that occur concurrently are possible.

2. All living organisms have certain basic requirements for their survival and wellbeing. Examples include the environmental conditions (e.g. temperature, pressure, pH and salinity) as well as food chains and the biogeochemical cycles.

3. A physical model (proposed by ESA) for CTPs is introduced.

4. Possible scenarios in the new climate equilibrium after 1.5°C have been explored using climate modelling. They include

   - The determination of 1.5°C as the warming temperature limit.

   - How an Earth System Model (ESM) and an appropriate ecological model may be used to model the impacts of coral bleaching events near 1.5°C and 1.8°C.

5. Four case studies of the actual (observed) climate impacts on humans and ecosystems are given, comprising

   - The 2003 Summer Heat Wave in Europe which is estimated to have caused 30,000 deaths.

   - The Heat Wave of July 2023 which struck South West USA and Mexico, Southern Europe and China.

   - The Global Coral Bleach Event which started in January 2023 and has exposed 84% of the world's reefs to high (bleaching level) temperatures.

   - The April 2024 Forest Ecosystem Collapse event that occurred in Western Australia.

6. The important threats due to the new climate conditions after 1.5°C are prioritised as extreme heat stress, food unavailability (famine)

and ecosystem collapse. A call for action on extreme heat stress was made by the Secretary-General of the UN in July 2024.

7. The following speculations are made on the worst possible future climate impacts:

- The likelihood of oxygen depletion due to AMOC collapse has been investigated, but there is no unanimous agreement. However, there is evidence of the loss of 0.5%–3.3% dissolved oxygen in the oceans during the 1970–2010 period.

- A simple statistical analysis shows that poorer countries (Angola, population 39 million) are much less capable of surviving extreme heat stress events than richer countries (Canada, population 41 million). An extreme heat stress event would put 12 million people in extreme danger of dying in Angola, while only 227,000 people would face the same risk in Canada.

- Biosphere collapse: the biosphere is like a global ecosystem. Examples of ecosystem collapse are already available, and the possibility of more severe cases in the future is real. The POVs for a total biosphere collapse can be the same as those for ecosystems, namely, the food chains and the biogeochemical cycles, providing the essential elements for the production of the four macromolecules of life.

The collapse can be triggered by extreme droughts/heat waves. Possible signals are the increasing frequencies of droughts and heat wave events.

REFERENCES

1. Copernicus. The 2024 annual climate summary. Global Climate Highlights 2024. 10 Jan 2025. Available from https://climate.copernicus.eu/global-climate-highlights-2024. Accessed 12 Feb 2025.
2. ESA. Understanding climate tipping points. Available from https://www.esa.int/Applications/Observing_the_Earth/Space_for_our_climate/Understanding_climate_tipping_points. Accessed 21 April 2025.
3. Wunderling, N et al. Climate tipping point interactions and cascades: A review. Earth System Dynamics 15 (2024) 41–74. Available from https://esd.copernicus.org/articles/15/41/2024/esd-15-41-2024.pdf. Accessed 11 Dec 2024.
4. Van der Gest, K., & Van den Berg, R. Slow onset events: A review of the evidence from the IPCC special reports on Land, Oceans and Cryosphere. Current Opinion in Environmental Sustainability 50 (June 2021) 109–120. Available from https://www.sciencedirect.com/science/article/pii/S1877343521000476. Accessed 24 April 2025.

5. Britannica. Environmental conditions. Available from https://www.britannica. com/science/biosphere/Environmental-conditions. Accessed 20 April 2025.

6. ELC. What gas do all animals need to survive?. Available from https:// enviroliteracy.org/what-gas-do-all-animals-need-to-survive/. Accessed 16 May 2025.

7. Singh, A. 2023. Bioenergy for Power Generation, Transportation and Climate Change Mitigation. IOP Publishing. Bristol. Available from https:// store.ioppublishing.org/page/detail/?kyt=climate%20generation%20 from&loc=uk. Accessed 20 April 2025.

8. Britannica. Environmental conditions. Available from https://www.britannica. com/science/biosphere/Environmental-conditions. Accessed 29 May 2025.

9. UCAR. Center for Science Education. Biogeochemical cycles. Available from https://scied.ucar.edu/learning-zone/earth-system/biogeochemical-cycles. Accessed 21 April 2025.

10. NCA. Report. Biogeochemical cycles. Available from https://nca2014.glo-balchange.gov/report/sectors/biogeochemical-cycles. Accessed 21 April 2025.

11. Special Report. Global warming of 1.5°C. Available from https://www.ipcc. ch/sr15/. Accessed 30 April 2025.

12. IPCC Sixth Assessment Report. Working group 1. The physical science basis. Available from https://www.ipcc.ch/report/ar6/wg1/. Accessed 2 May 2025.

13. IPCC Sixth Assessment Report. The physical science basis. Summary for policymakers. Available from https://www.ipcc.ch/report/ar6/wg1/chapter/ summary-for-policymakers/. Accessed 6 May 2025.

14. Frieler et al. Limiting global warming to 2°C is unlikely to save most coral reefs. Nature Climate Change 3 (2013) 165–170. Available from https:// www.nature.com/articles/nclimate1674. Accessed 29 April 2025.

15. UNEP. Environment alert bulletin. Impacts of summer 2003 heat wave in Europe. Available from https://www.unisdr.org/files/1145_ewheatwave. en.pdf. Accessed 8 May 2025.

16. World Weather Attribution. Extreme heat in North America, Europe and China in July 2023 made much more likely by climate change. 25 July 2023. Available from https://www.worldweatherattribution.org/extreme-heat-in-north-america-europe-and-china-in-july-2023-made-much-more-likely-by-climate-change/. Accessed 9 May 2025.

17. The Guardian. Climate crisis. More than 80% of the world's reefs hit by bleaching after worst global event on record. Graham Readfern. Wed 23 April 2025. Available from https://www.theguardian.com/environment/2025/ apr/23/coral-reef-bleaching-worst-global-event-on-record?utm_ source=cbnewsletter&utm_medium=email&utm_term=2025-05-04&utm_campaign=DeBriefed±Brazil±calls±for±country±emissions±pla ns±Global±coral±bleaching±Where±top±pope±contenders±stand±on±cli mate. Accessed 10 May 2025.

18. Allen, C. D. et al. A global overview of drought and heat-induced tree mortality reveals emerging climate change risks for forests. Forest Ecology and Management 259 (4) (2010) 660–684.

19. ABC News. Fears of another "forest collapse" event in Western Australia after record dry spell. Briana Shepherd. 11 April 2024. Available from https://www.abc.net.au/news/2024-04-11/ecologists-warn-potential-forest-collapse-event-wa/103682304. Accessed 11 May 2025.

20. Australian Geographic. 19 Australian ecosystems are already collapsing: Which one will fall first? 26 February 2021. Available from https://www.australiangeographic.com.au/news/2021/02/19-australian-ecosystems-are-already-collapsing-which-one-will-fall-first/. Accessed 11 May 2025.

21. IPCC. About the IPCC. Available from https://www.ipcc.ch/about. Accessed 14 May 2025.

22. United Nations Climate Action. Secretary-General's call to action on extreme heat. Available from https://www.un.org/en/climatechange/extreme-heat. Accessed 16 May 2025.

23. United Nations. Secretary-General's call to action on extreme heat. 25 July 2024. https://www.un.org/sites/un2.un.org/files/unsg_call_to_action_on_extreme_heat_for_release.pdf. Accessed 16 May 2025.

24. NOAA. What is the Atlantic meridional overturning circulation? Available from https://oceanservice.noaa.gov/facts/amoc.html. Accessed 17 May 2025.

25. AtlanticOcean.info. 2 February 2025. How changes in the Atlantic's thermohaline circulation could impact global weather. Available from https://atlanticocean.info/2025/02/02/how-changes-in-the-atlantics-thermohaline-circulation-could-impact-global-weather/. Accessed 17 May 2025.

26. The Journal. Climate change: Amocalypse now. December 2024/January 2025. Available from https://thejournal.cii.co.uk/2025/01/10/climate-change-amocalypse-now. Accessed 17 May 2025.

27. Nature Environment. 27 March 2024. One of the World's most important ocean currents really is slowing down. It's what oceanographers have feared, but had difficulty proving. Available from https://www.iflscience.com/one-of-the-worlds-most-important-ocean-currents-really-is-slowing-down-73554. Accessed 17 May 2025.

28. AR6 WG1 Report. Chapter 5. Available from https://www.ipcc.ch/report/ar6/wg1/chapter/chapter-5/. Accessed 17 May 2025.

29. Salin et al. Banque de France. Working paper. Assessing integrated assessment models for building nature-economy scenarios. Available from https://www.banque-france.fr/system/files/2024-08/WP959.pdf. Accessed 29 May 2025.

30. UN Climate Change. IAMs and energy-environment-economy models. Available from https://unfccc.int/topics/mitigation/workstreams/response-measures/modelling-tools-to-assess-the-impact-of-the-implementation-of-response-measures/integrated-assessment-models-iams-and-energy-environment-economy-e3-models#GEM-E3. Accessed 29 May 2025.

31. UN-Habitat. Urban indicators database. Housing, slums and informal settlements. Available from https://data.unhabitat.org/pages/housing-slums-and-informal-settlements. Accessed 29 May 2025.

32. Tracking SDG7. The energy progress report 2024. Available from https://trackingsdg7.esmap.org/data/files/download-documents/sdg7-report2024-0611-v9-highresforweb.pdf. Accessed 20 May 2025.

33. Britannica. Biosphere. Available from https://www.britannica.com/science/biosphere. Accessed 22 May 2025.

34. Organismal Biology. A brief history of geologic time. Available from https://organismalbio.biosci.gatech.edu/biodiversity/prokaryotes-bacteria-archaea-2/#:~:text=The%20fossil%20record%20indicates%20that%20the%20first%20living,4.6%20billion%20years%20old%20based%20on%20radiometric%20dating. Accessed 22 May 2025.

35. National Geographic. Biosphere. Available from https://education.nationalgeographic.org/resource/biosphere/ Accessed 22 May 2025.

36. Outforia. Types of ecosystems. Available from https://outforia.com/types-of-ecosystems/. Accessed 23 May 2025.

37. Harvest Harmonies Understanding Ecosystems in Agriculture. By Dr. Ramesh Babu. Available from https://harvestharmonies.com/articles/understanding-ecosystems-agriculture/. Accessed 23 May 2025.

38. National Geographic. Ecosystem. Available from https://education.nationalgeographic.org/resource/ecosystem. Accessed 23 May 2025.

39. PMF IAS. Energy flow through an ecosystem: food chain, food web. 31 Dec 2024. Available from https://www.pmfias.com/trophic-levels-food-chain-food-web-biotic-interaction/. Accessed 26 May 2025.

40. Britannica. The importance of the biosphere. Available from https://www.britannica.com/science/biosphere/The-importance-of-the-biosphere. Accessed 26 May 2025.

41. Beck, K. What are the four macromolecules of life? Sciencing. 4 March 2022. Available from https://www.sciencing.com/four-macromolecules-life-8370738/. Accessed 27 March 2025.

42. USGC. Climate-induced forest dieback: An escalating global phenomenon? Avail from https://pubs.usgs.gov/publication/70036616. Accessed 28 Mar 2025.

43. FAO. Small family farmers produce one-third of the world's food. Available from https://www.fao.org/newsroom/detail/Small-family-farmers-produce-a-third-of-the-world-s-food/en. Accessed 29 May 2025.

44. ETC Group. Small-scale farmers and peasants still feed the world. Available from https://www.etcgroup.org/files/files/31-01-2022_small-scale_farmers_and_peasants_still_feed_the_world.pdf. Accessed 29 May 2025.

CHAPTER 9

New Thinking on a More Viable Strategy

9.1 INTRODUCTION

The aim of this chapter is to propose a new climate action framework that views the climate crisis from a broader perspective. This new framework incorporates actions and considerations from other domains (spheres of action and interest) and provides a new basis for an all-inclusive climate action strategy for achieving the climate goals.

Chapters 1 and 2 noted that the current methods of addressing the climate challenge were not adequate, and that new thinking was needed to re-frame the seemingly intractable problem of the climate crisis. We are now in a position to address this issue.

In Section 6.4, a case was made for a formal development approach to a country's net-zero strategy, and the *Systems Development Life Cycle* approach was identified as a possible contender. This life cycle approach also provides insights for the design of the new solution being sought.

In Section 7.6, it was pointed out that

- there were other methods as well (apart from emissions reduction) that could contribute to limiting warming to 1.5°C, and that

- the implementation of climate strategies could be influenced by political decision-making, geopolitical events and actions in the economic domain, all of which had to be taken into account.

DOI: 10.1201/9781003531180-9

All these observations point to a new strategy that

- is focused on limiting global warming, rather than being confined to emissions reductions alone,

- benefits from the ever-growing volume of climate science research,

- widens the solution scope of the strategy, and

- consolidates actions from the economic, physical, political and geopolitical domains into the scientific domain where the current net-zero strategies largely reside.

A start to formulating such a strategy can be made by emulating the *Systems Development Life Cycle methodology* invoked for net-zero strategies in Chapter 6. The events that have unfolded over the last five years have revealed the need for a more versatile method that accommodates actions from several domains. Such a *Quasi-Life Cycle Development methodology* will be capable of accommodating new learnings and re-learnings, as well as consolidating climate-related actions and influences from other domains.

This chapter begins by considering such a strategy, put together through a simple process of concatenation. It then proposes a more elegant representation in the form of the *rationalised net-zero strategy (RNZS)*. The chapter ends by noting that the nature of the problem itself has changed in the course of recent events, and the desired solution must be versatile and resilient enough to accommodate to such new conditions.

9.2 THE NEW FRAMEWORK

The need for a broader perspective and the inclusion of extraneous influences on climate action over the last few years calls for a new approach to the framing and development of climate action plans. These two requirements are needed to

- shift the focus to the original cause of climate change, and to

- include climate-related actions/events from other domains (some of which have emerged only recently) that influence the success of the climate strategy.

To ensure success, such influences will need to be fully integrated into the action plan. This section takes a closer look at what these influences are, beginning with an examination of the physical basis for the cause of climate change.

9.2.1 Broadening the Physical Basis

Confining the climate action strategy to one of reducing emissions over-looks the key cause of climate change. This is global warming, which occurs because the Earth's energy balance is disturbed.

In Chapter 3, the Earth's energy balance was defined as the condition where the net incoming radiation from the sun became equal to the net outgoing thermal radiation from the Earth. This balance is usually evaluated at the Top of the Atmosphere (TOA). One can divide the global warming process into two parts, the first where there are no GHGs present in the atmosphere, and the second where a fixed amount of GHG is injected into the Earth's atmosphere.

With no GHGs in the atmosphere, the net incoming (short wavelength) radiation from the sun becomes equal to the net outgoing (long wavelength) radiation (OLR) from the Earth at equilibrium. Referring to NASA's Earth Energy Budget shown in Figure 3.1, and inspecting the incoming and outgoing radiation, it is seen that the net incoming radiation from the sun is simply the total absorbed solar radiation (ASR). This remains essentially constant, and at the TOA, this is equal to the OLR at equilibrium.

When the GHGs are introduced into the atmosphere, some of the outgoing long wavelength radiation is radiated back by the GHG. This reduces the OLR by sending some of it back towards the Earth. Because of this extra radiation received, the Earth begins warming again till a new balance (ASR=OLR) is achieved. This second warming is the global warming that is of concern to us.

We can reduce the global warming by reducing the concentration of GHGs in the atmosphere. But there are also other ways of reducing warming. Two such ways are to

- let more incoming (short-wavelength) radiation be reflected back (see Figure 3.1). This reduces the ASR at TOA, so less solar radiation (which is the original source of the Earth's warming) is received by the Earth, or

- increase the net outgoing long wavelength radiation (this increases the OLR).

These are two purely radiative measures that can be taken to reduce global warming. These measures do not appear in any net-zero strategy considered by the Parties of the Convention currently. The reason is simple: net-zero strategies only deal with controlling the amount of GHG emissions into the atmosphere and thus are only concerned with the concentration of GHGs present in the atmosphere.

Thus, it is clear that, in constructing net-zero strategies, starting from a consideration of the energy balance (and global warming) will introduce two other ways of dealing with the problem, i.e. it will increase the size of the solution set. This clearly opens up more options for possible net-zero strategies as compared to those available earlier. It must be stressed that, while geo-engineering is one such possible solution, this is not being advocated here, as there are other options available for producing the changes in the incoming and outgoing radiation suggested above.

In addition to broadening the physical basis relevant to climate change mitigation, there are also elements from other domains that influence the success of a viable climate action strategy. These are discussed in the subsections below.

9.2.2 Economic Actions: Positive Tipping Points

How actions in the socio-economic domain can play a determining role in the climate action strategy was first mentioned in Chapter 4, where it was noted that such actions could create positive tipping points (PTPs) that could expedite solutions to the net-zero challenge [1, 2].

The authors of the *Global Tipping Point Report 2023* [1] describe PTPs by noting that they occur when the balance of system feedbacks shifts in favour of reinforcing ones, and give economies of scale and social contagion as examples.

They also note that

> PTPs are already well underway in wind and solar power generation and in leading battery electric vehicle (BEV) markets.

One of the challenges that needs to be surmounted by a net-zero strategy is to accelerate the rate of renewable energy technology uptake to meet the 2050 requirement. It is well understood in economics that this can be achieved by reducing the cost of producing the technology. As noted by Griffith [3], two ways for reducing the cost of any energy technology are *Learning by researching* and *Learning by Doing*. In the first, improvements are made in the technology itself through research. The latter method drives down the costs of production through the learning and experience gained as the cumulative volume of the technology produced increases.

The process of cost reduction in the latter method is described by the *Learning Curve* for the technology, and the relation between the price reduction and the cumulative production is called *Wright's Law*, discovered by T. P. Wright for airplanes in 1936 [4]. The learning curve of a particular technology is characterised by its *Learning Rate*, which is defined as *the fraction (%) by which the cost of the technology reduces with a doubling of its global installed capacity* (see Table 7.1 given by Singh [5]).

A PTP for a technology is crossed in the economic domain when its cost of production (and consequently its market price) begins to tumble in the market, thus accelerating the uptake of the technology. When the production of two or more technologies is coupled somehow (as in the case of the production of battery-driven EVs and their batteries), an acceleration in the production of one leads to a similar acceleration in the other, and a cascade follows.

The formation of such cascades in the socio-economic domain provides the central principle advocated for climate action by Lenton and his group in the GTP Report [1].

Note that a PTP has a positive effect on climate action strategy.

9.2.3 Political and Geo-Political Shifts

The other actions/events that have taken place in the physical, political and geo-political domains recently that are of consequence are COVID-19, the wars in Ukraine and Gaza, as well as the start of President Trump's Make America Great Again (MAGA) campaign.

The former three events impact climate action largely by detracting attention from the climate crisis (i.e. lower its priority status) and are expected to have significant detrimental influence on the success of climate action such as causing delays in the action plan. In comparison, Trump's MAGA campaign, which involves withdrawal of support from the Paris Agreement and actively increasing GHG emissions, is in direct opposition to the climate action strategy and will have a significant negative impact on it.

Note that all these economic and political/geo-political events have previously been treated as factors that were external to the climate solution being developed. To take account of their influence, these need to be internalised into the mainstream of the action strategy.

9.3 THE QUASI-LIFE CYCLE APPROACH TO DEVELOPING STRATEGIES

In the last section, we elaborated on the factors from other domains that influenced the success of the climate action strategy. It was seen that this impact could be positive, detrimental or negative. The new strategy begins by acknowledging that these are part of the problem, and incorporating them into the strategy to "internalise" them.

How is this action executed?

The answer lies in the manner in which the action strategy is developed. It was recommended in Chapter 6 that a formal procedure be used for developing net-zero strategies, and a development cycle approach was suggested. The new climate strategy for limiting the Earth's warming may be obtained through a development cycle similar to that for the net-zero strategies considered in Chapter 6. However, it must now be applied to the new framework developed in this chapter. As this cycle will contain new actions that were not present in the earlier cycle, it is a modified version of the first. The development cycle that results is the *quasi-life cycle development methodology*. This is depicted in Figure 9.1.

The stages of this development cycle are described in Table 9.1.

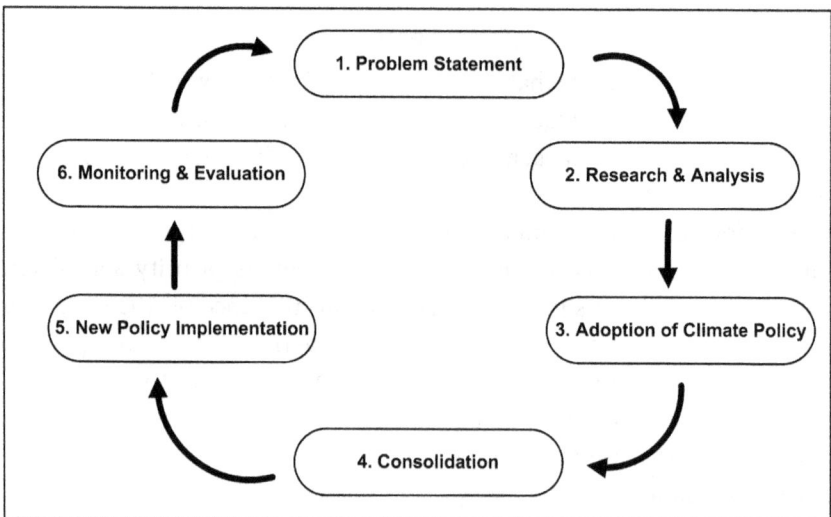

FIGURE 9.1 The quasi-life cycle approach to the development of the new strategy. (See Table 9.1 for descriptions of the stages). (Figure credit: Asha Sinha.)

TABLE 9.1   Stages of the Quasi-Life Cycle Development Cycle (Q-LCDC)

Stage	Description
1. Problem statement	Define the objectives and scope of the climate strategy to be developed.
2. Research and analysis	Essential and desirable criteria, and the current status of science for a viable climate strategy
3. Adoption of climate policy	The Paris Agreement
4. Consolidation	Assimilate all the influences listed in the new framework
5. New policy implementation	Implement this new (consolidated) policy to produce the new (concatenated) climate strategy
6. Monitoring and evaluation	Monitor the strategy in operation and note any deficiencies/recommendations

As can be seen from the figure and the accompanying table, the new development cycle is different from the earlier one (see Figure 6.1) in the following stages:

- Stage 2: The requirements analysis has been replaced by a stage labelled Research & Analysis. This accommodates the new learnings acquired through recent climate science research.

- Stage 3: The design stage here has been replaced by the Adoption of a Climate Policy. This policy is the Paris Agreement itself.

- Stage 4: The development stage here has been replaced by a Consolidation stage. In this stage of the cycle, all the new actions/events from the other domains (that were not mentioned in the previous strategy) are added to the new strategy through a simple concatenation process.

- Stage 5: The new policy that results from the concatenation process in the previous stage is put into action.

As mentioned before, the new strategy that results from the development process above is a concatenated strategy, i.e. one that includes elements that are simply added on. How these elements can be assimilated in a more seamless manner is the subject of the next section.

9.4 THE RATIONALISED NET-ZERO STRATEGY (RNZS)

An issue with the new (concatenated) strategy is that while it achieves its aim of internalising actions from all domains within the same strategy structure, it does not explicitly express the nature and origins of the new inclusions, or how the strategy will function as a whole. One therefore

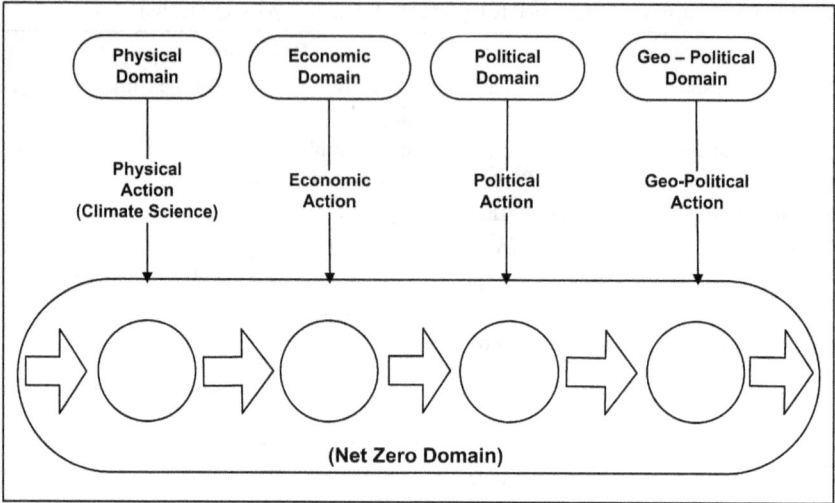

FIGURE 9.2 A domain-mapping diagram showing how information flow takes place in the rationalised net-zero strategy (RNZS). (Figure credit: Asha Sinha.)

needs to embed it within an encompassing structure that demonstrates, in a more elegant manner, how the actions from other domains are to be included seamlessly into the mainstream net-zero strategy.

The *RNZS* does this using a *domain mapping procedure* that provides a suitable mechanism for visualising the manner in which information flows take place within the RNZS. This domain mapping is shown in Figure 9.2.

In Figure 9.2, the mainstream net-zero strategy exists in the net-zero domain as shown. The actions introduced through the new framework exist in their respective domains at the top of the figure, and are mapped onto the net-zero domain at the circles, which represent the *information interchange hubs* where information is exchanged between the two domains.

The RNZS is a radically different strategy compared to the previous climate action strategies. It is designed to accept change and to adapt to it. Careful examination of Figure 9.2 should demonstrate that

The RNZS is a living strategy that grows in an evolutionary way with the changing situation.

Unlike previous strategies which are unchanging and inert, it is a dynamic action plan that responds to external shocks by pausing and re-adjusting itself to the new demand.

It anticipates change and provides mechanisms for dealing with it.

9.5 CONCLUDING REMARKS

This search for a viable solution to the climate crisis began at a time when the world was in the grips of a multitude of crises. It was expected that things would turn for the better. But a year later, the situation has taken a turn for the worse, with a change in the global order that is threatening the very existence of democracy as we have known it in recent times.

The global ethos has undergone a dramatic transition, from one where nations were predominantly strategising for the good of all through programmes such as the MDGs and the SDGs, to one where strategising for their own national security has become their major concern. Under such circumstances, climate action has assumed a priority status that is way below those of national defence and security.

Meanwhile, scientific data on the status of the climate, as reported by leading climate authorities, continue to show that we are still headed for the type of "climate disaster" that was the subject of deliberation in the last chapter. The question that looms large is whether it will lead to a physical collapse of the climate system resulting in socio-economic mayhem.

Previous human experience tells us that this will probably not happen. This is because of the extraordinary capacity that humans have displayed in the past to survive at times of extreme crises through the remarkable resourcefulness and resilience that always surfaces under such circumstances.

However, in all probability, only the richest nations will be a position to summon the resources for the relevant solutions. This will leave the less fortunate nations to face the full brunt of the crisis.

Thus, the first indications of the likely future outcomes of the climate crisis seem to be a widening of the gap between the rich nations and the poor.

As mentioned above, the requirements of global/national security are now receiving higher priority than the climate crisis. The hope is that the world will revert to the "peace-time" mode soon. Maybe it is a time to wait. Such a decision is indeed commensurate with the RNZS.

9.6 SUMMARY

1. A new climate action framework is presented that views climate action from a broader perspective.

2. In addition to containing and reducing the GHG concentrations in the Earth's atmosphere, global warming can also be limited by

reducing the net incoming (short wavelength) solar radiation to the Earth, or by increasing the net outgoing (long wavelength) radiation from the Earth. A net-zero strategy should also include these additional options.

3. Actions in the socio-economic domain can assist in accelerating the transition to renewable energy through the creation of cascading PTPs in the economic domain. Policies and actions in political and geo-political spheres can have detrimental or negative impacts on climate action strategies. All such actions, policies and events, which were previously being treated as being external to the strategy, should be included as part of the mainstream action of the strategy.

4. The external influences on net-zero strategies can be included into the new strategy via a life cycle development methodology similar to that proposed in Chapter 6. This new development methodology is called the *quasi life cycle development methodology* to account for its new structure.

5. It must be noted that the new strategy is a concatenated strategy, i.e. is formed simply be adding on elements to the old strategy. It does not show how elements from the other domains will be included in the new strategy.

6. The *RNZS* uses a domain-mapping approach to show how actions/events from other domains will be included during the use of the strategy. It is a living strategy that grows in an evolutionary way with the changing situation.

REFERENCES

1. Lenton, T. M. et al. (eds.) 2023. The Global Tipping Points Report 2023. University of Exeter, Exeter, UK. Available from https://global-tipping-points.org/resources-gtp/. Accessed 11 Dec 2024.
2. World Economic Forum. Climate action. Positive tipping points: A credible way to meet climate and nature goals. 8 July 2024. Available from https://www.weforum.org/stories/2024/07/positive-tipping-points-climate-nature-goals-wef/. Accessed 11 Dec 2024.
3. Griffith, S. 2022. The Big Switch, Australia's Electric Future. Black Inc., Collingwood, Australia.

4. Wright, T. P. Factors affecting the cost of airplanes. Journal of the Aeronautical Sciences 3 (1936) 122.
5. Talking Renewables (Second Edition). A renewable energy primer for everyone. March 2025. IOP Publishing. Available from https://store.ioppublishing.org/page/detail/Talking-Renewables-Second-Edition/?K=9780750362788. Accessed 24 May 2025.

Index

Note: Page references with *Italics* refer to figures, **bold** refer to tables.

For Product Safety Concerns and Information please contact our EU
representative GPSR@taylorandfrancis.com
Taylor & Francis Verlag GmbH, Kaufingerstraße 24, 80331 München, Germany

www.ingramcontent.com/pod-product-compliance
Lightning Source LLC
Chambersburg PA
CBHW070723220326
41598CB00024BA/3267